Mesh Adaptation for Computational Fluid Dynamics 2

Mesh Adaptation for Computational Fluid Dynamics 2

Unsteady and Goal-oriented Adaptation

Alain Dervieux
Frédéric Alauzet
Adrien Loseille
Bruno Koobus

WILEY

First published 2022 in Great Britain and the United States by ISTE Ltd and John Wiley & Sons, Inc.

Apart from any fair dealing for the purposes of research or private study, or criticism or review, as permitted under the Copyright, Designs and Patents Act 1988, this publication may only be reproduced, stored or transmitted, in any form or by any means, with the prior permission in writing of the publishers, or in the case of reprographic reproduction in accordance with the terms and licenses issued by the CLA. Enquiries concerning reproduction outside these terms should be sent to the publishers at the undermentioned address:

ISTE Ltd
27-37 St George's Road
London SW19 4EU
UK

www.iste.co.uk

John Wiley & Sons, Inc.
111 River Street
Hoboken, NJ 07030
USA

www.wiley.com

© ISTE Ltd 2022

The rights of Alain Dervieux, Frédéric Alauzet, Adrien Loseille and Bruno Koobus to be identified as the authors of this work have been asserted by them in accordance with the Copyright, Designs and Patents Act 1988.

Any opinions, findings, and conclusions or recommendations expressed in this material are those of the author(s), contributor(s) or editor(s) and do not necessarily reflect the views of ISTE Group.

Library of Congress Control Number: 2022934929

British Library Cataloguing-in-Publication Data
A CIP record for this book is available from the British Library
ISBN 978-1-78630-832-0

Contents

Acknowledgments ix

Introduction xi

Chapter 1. Nonlinear Corrector for CFD 1

  1.1. Introduction 1
      1.1.1. Linear correction 3
      1.1.2. Nonlinear correction 4
  1.2. Two correctors for the Poisson problem 5
      1.2.1. Notations 5
      1.2.2. A priori corrector for the PDE solution 6
      1.2.3. Finer-grid DC corrector for the PDE solution ... 8
  1.3. RANS equations 9
      1.3.1. Vector form of the RANS system 9
      1.3.2. Formal discretization 10
      1.3.3. Notations for discretization 11
  1.4. Nonlinear functional correction 13
      1.4.1. Finite volume nonlinear corrector 13
      1.4.2. Finite element corrector 15
  1.5. Example: supersonic flow 17
  1.6. Concluding remarks 18
  1.7. Notes 20

Chapter 2. Multi-scale Adaptation for Unsteady Flows 21

  2.1. Introduction 21
  2.2. Mesh adaptation efficiency 23
      2.2.1. Regular and singular unsteady model 23
      2.2.2. Representativity of the spatial interpolation error ... 24
  2.3. Transient fixed-point mesh adaptation scheme 25

2.3.1. Size of subintervals in a mesh convergence 28
2.3.2. Mesh adaptation for unsteady Euler/Navier–Stokes equations with thickened interface 29
2.3.3. Convergent transient fixed-point 33
2.4. 2D bi-fluid example 33
2.5. Example: impact of a 3D water column on a obstacle 35
2.6. Conclusion 39
2.7. Appendix: remarks about the adaptation of the time step 39
2.8. Notes 41

Chapter 3. Multi-rate Time Advancing 43

3.1. Introduction 43
3.2. Multi-rate time advancing by volume agglomeration 45
3.2.1. Finite volume Navier–Stokes 45
3.2.2. Inner and outer zones 46
3.2.3. MR time advancing 47
3.3. Elements of analysis 49
3.3.1. Stability 49
3.3.2. Accuracy 50
3.3.3. Efficiency 51
3.3.4. Toward many rates 52
3.3.5. Impact of our MR complexity on mesh adaption 52
3.3.6. Parallelism 53
3.4. Applications 55
3.4.1. Circular cylinder at very high Reynolds number 55
3.4.2. Mesh adaption for a contact discontinuity 58
3.5. Conclusion 59
3.6. Notes 60

Chapter 4. Goal-Oriented Adaptation for Inviscid Steady Flows . . . 65

4.1. Introduction 65
4.1.1. What to do with this estimate? 67
4.1.2. Adjoint-L^1 approach 68
4.1.3. Outline 69
4.2. A more accurate nonlinear error analysis 69
4.2.1. Assumptions and definitions 69
4.2.2. A priori estimation 70
4.3. The case of the steady Euler equations 72
4.3.1. Variational analysis 72
4.3.2. Approximation error estimation 73
4.4. Error model minimization 74
4.5. Adaptative strategy 76
4.5.1. Adjoint solver 77

4.5.2. Optimal goal-oriented discrete metric 77
4.5.3. Controlled mesh regeneration 79
4.6. Numerical outputs 79
   4.6.1. High-fidelity pressure prediction of an aircraft 79
4.7. Conclusion 82
4.8. Notes 82

Chapter 5. Goal-Oriented Adaptation for Viscous Steady Flows . . 85

5.1. Introduction 85
5.2. Case of an elliptic problem 86
   5.2.1. A priori finite-element analysis (first estimate) 86
   5.2.2. Goal-oriented adaptation according to lemma 5.1 89
   5.2.3. Goal-oriented adaptation according to a second estimate 91
5.3. Error analysis for Navier–Stokes problem 92
   5.3.1. Mesh adaptation problem statement 92
   5.3.2. Linearized error system 93
   5.3.3. First estimate for Navier–Stokes problem 94
   5.3.4. Second estimate for Navier–Stokes problem 98
   5.3.5. Optimal goal-oriented continuous mesh 101
5.4. From theory to practice 101
   5.4.1. Computation of the optimal continuous mesh 103
5.5. An example of application to a turbulent flow 103
5.6. Conclusion 107
5.7. Notes 109

Chapter 6. Norm-Oriented Formulations 111

6.1. Introduction 111
6.2. A summary of previous analyses 114
   6.2.1. Feature-based adaptation by interpolation error optimization 114
   6.2.2. Implicit a priori error estimate and corrector 115
   6.2.3. Goal-oriented analysis 116
6.3. Norm-oriented approach 118
6.4. Numerical elliptic examples 119
   6.4.1. Numerical features 119
   6.4.2. 2D boundary layer 122
   6.4.3. Poisson problem with discontinuous coefficient 123
6.5. Application to flows 126
   6.5.1. A comparison feature-oriented/norm 127
   6.5.2. Application to a viscous flow 129
6.6. Conclusion 130
6.7. Notes 131

viii   Mesh Adaptation for Computational Fluid Dynamics 2

Chapter 7. Goal-Oriented Adaptation for Unsteady Flows 133

   7.1. Introduction 133
   7.2. Formal error analysis 134
   7.3. Unsteady Euler models 135
      7.3.1. Continuous state system and finite volume formulation 135
      7.3.2. Continuous adjoint system and discretization 137
      7.3.3. Impact of the adjoint: numerical example 141
   7.4. Optimal unsteady adjoint-based metric 142
      7.4.1. Error analysis for the unsteady Euler model 142
      7.4.2. Continuous error model 144
      7.4.3. Spatial minimization for a fixed t 146
      7.4.4. Temporal minimization 146
      7.4.5. Temporal minimization for time sub-intervals 150
   7.5. Theoretical mesh convergence analysis 155
      7.5.1. Smooth flow fields 155
   7.6. From theory to practice 157
      7.6.1. Choice of the GO metric 158
      7.6.2. Global fixed-point mesh adaptation algorithm 158
      7.6.3. Computing the GO metric 161
   7.7. Numerical experiments 161
      7.7.1. 2D Acoustic wave propagation 161
      7.7.2. 3D blast wave propagation 163
   7.8. Conclusion 165
   7.9. Notes 166

Chapter 8. Third-Order Unsteady Adaptation 167

   8.1. Introduction 167
   8.2. Higher order interpolation and reconstruction 168
   8.3. CENO approximation for the 2D Euler equations 170
      8.3.1. Model 170
      8.3.2. CENO formulation 171
      8.3.3. Vertex-centered low dissipation CENO2 174
   8.4. Error analysis 175
   8.5. Metric-based error estimate 178
   8.6. Optimal metric 179
   8.7. From theory to practical application 182
   8.8. A numerical example: acoustic wave 183
   8.9. Conclusion 186
   8.10. Notes 186

References 189

Index 199

Summary of Volume 1 201

Acknowledgments

This book presents many theoretical and numerical accomplishments performed in collaboration with the following researchers:

Rémi Abgrall, Olivier Allain, Francoise Angrand, Paul Arminjon, Nicolas Barral, Anca Belme, Fayssal Benkhaldoun, Francois Beux, Gautier Brèthes, Véronique Billey, Alexandre Carabias, Romuald Carpentier, Giles Carré, Yves Coudière, Francois Courty, Didier Chargy, Paul-Henri Cournède, Christophe Debiez, Jean-Antoine Desideri, Gérard Fernandez, Loula Fezoui, Jérôme Francescatto, Loic Frazza, Pascal Frey, Paul-Louis George, Aurélien Goudjo, Nicolas Gourvitch, Damien Guégan, Hervé Guillard, Emmanuelle Itam, Marie-Hélène Lallemand, Stéphane Lanteri, Bernard Larrouturou, Anne-Cécile Lesage, David Leservoisier, Francoise Loriot, Mark Loriot, Laurent Loth, Nathalie Marco, Katherine Mer, Victorien Menier, Bijan Mohammadi, Eric Morano, Boniface Nkonga, Géraldine Olivier, Bernadette Palmerio, Gilbert Rogé, Bastien Sauvage, Éric Schall, Hervé Stève, Bruno Stoufflet, Francois Thomasset, Julien Vanharen, Ganesan Vijayasundaram, Cécile Viozat and Stephen Wornom; we also want to apologize to the people we forgot to mention.

Also we want to acknowledge our friends of INRIA and Lemma, and in particular Charles Leca, Olivier Allain, Nathalie and Philippe Boh, for their support. INRIA provided excellent conditions for research and writing of this book to the first three authors. Lemma permitted a rapid industrialization of our mesh adaptation methods.

The first author thanks his advisers, Jean Céa, Roland Glowinski and many thanks also to Charbel Farhat, Jacques Périaux and Roger Peyret.

This study is supported by fp6 and fp7 European progams (AEROSHAPE, HISAC, NODESIM, UMRIDA). The authors and their coworkers were granted access to the HPC resources of CINES/IDRIS under allocations made by GENCI (Grand Equipement National de Calcul Intensif).

Introduction

Numerical simulation is a central tool in the design of new human artifacts. This is particularly true in the present decades due to the difficult challenge of climate evolution. Yet recently climatic constraints were simply translated into the need for further progress in reducing pollution, a big job, in particular for specialists of numerical simulation. Today, it is likely that the use of numerical simulation, and particularly computational mechanics, will be central to the study of a new generation of human artefacts related to energy and transport. Fortunately, these new constraints are contemporary with the rise of a remarkable maturity of numerical simulation methods. One sign of this maturity is the flourishing of mesh adaptation. Indeed, mesh adaptation is now able to solve in a seamless way the deviation between theoretical physics and numerical physics, managed by the computer after discretization. A practical manifestation of this is that the engineer is freed from taking care of the mesh(es) needed for analysis and design. A second effect of mesh adaptation is an important reduction of energy consumption in computations, which will be amplified by the use of so-called higher order approximations. Mesh adaptation is thus the source of a new generation of more powerful numerical tools. This revolution will affect a generation of conceptualizers, numerical analysts and users, who are the engineers in design teams.

These books (Volumes 1 and 2) will be useful for researchers and engineers who work in computational mechanics, who deal with continuous media, and in particular who focus on computational fluid dynamics (CFD). They present novel mesh adaptation and mesh convergence methods developed over the last two decades, in part by the authors. They are expanded from a series of scientific articles, which are re-written, reorganized and completed in order to make the new content up-to-date, self-contained and more educational.

Let us describe in our own way the central role of meshes in the numerical simulation process, making it possible to compute a prediction of a physical phenomenon. In short, real-life mechanics consists of molecules and their interactions. The notion of continuous medium helps to transform a large but finitely complex system into a infinitely but smoothly complex system. For example, the understanding of gas flow relies today on the kinetic gas theory, which says that a gas is made up of a large number of molecules. We imagine the molecules as balls (monoatomic gas), but this is only a model for our imagination. We next consider that these balls are playing some sort of 3D "billiard" and that, if we are lucky, a continuum model describing the macroscopic behavior is a representative mean of the individual behaviors. In a similar manner, solids consist of a large number of molecules interacting with each other and can be modeled as continuum media. Then the history of our billions of molecules is transformed into a continuous medium. Strictly speaking, the amount of information has gone from very large to infinitely large! But we have an extra assumption: the infinitely complex functions that describe the continuous medium are smooth almost everywhere because there exists a small scale such that even smaller scales behave in an expected way (predictable by interpolation, for example) except for some error that is very small.

This assumption allows us many mathematical strategies:

– some laws for such functions can be constructed by considering that the functions have (regular enough) derivatives, with respect to time, and/or to space;

– such functions can be accurately represented by a set of special functions described by a small amount of information, such as polynomials, thanks, typically, to the Taylor formula. The first point corresponds to the construction of partial differential equations (PDEs). The second point corresponds the approximation or interpolation of known functions or the approximation of PDE solutions.

For clarity, we shall speak about "interpolation" for a discrete representation close to a known function and about "approximation" only for a discrete representation close to an (a priori unknown) PDE solution. Both strategies, interpolation and approximation, are related to discretization, the purpose of which is to reduce the infinitely complex continuous model into finitely complex discrete models, so that both the representation and the seeking of these for a particular physical situation is the matter of a finite amount of computational resources:

– a finite number of digits;

– a finite number of operations.

With respect to digits, any real number can still be extremely/infinitely complex and has to be replaced by a floating point representation. We shall not analyze the difference between a real number and its floating point representations, which is out of the scope of this book. For us, the only consequences of replacing real numbers by floating point representations is that using round-off truncation may amplify error modes in iterative algorithms. We shall have to choose only numerical algorithms that will produce good results even with round-off error, namely stable algorithms.

With respect to operations, our discrete model will rarely be computed on a sheet of paper but more frequently with a computer. It is common to call the complexity of an algorithm the number of elementary operations (additions, multiplications, etc.) necessary to perform it on a given set of data. We extend this notion by defining the complexity of the interpolation of a function as the number of numbers necessary to represent (with a given accuracy) that function. Two important ingredients of numerical analysis are connected by this notion:

– to *approximate a known function* with a certain level of accuracy, we need to handle (to store) some amount of real numbers, the degrees of freedom;

– to *compute* these degrees of freedom for an unknown function using an algorithm solving a system to provide the approximate solution, we need accordingly some amount of operations.

In both cases, we need to adapt a data structure to the geometrical domain on which the function has its definition set. This data structure, which we call a mesh, locates nodes on the domain. Approximating a smooth function can be done with a storage that increases at best only inverse-exponentially (spectral approximation) or, in most cases, inverse polynomially (approximation of given order) with the prescribed error. The computation of this approximation can, at best, be done with linearly complex algorithms such as full multigrid. More precisely, let us consider the combination of

– a smooth unknown solution u of a PDE in a domain inside $I\!R^d$;

– an approximation of the PDE of order α for a certain norm $|.|$, that is,

$$|u - u_N| \leq K_1 N^{-\alpha/d}$$

with K_1 depending on the solution u and where N is the number of nodes of the computational domain; and

– a linearly complex solution algorithm, $CPU(N) = K_2 N$ with K_2 depending on the algorithm and where "CPU" is the computational time.

Then for a prescribed error $|u - u_N| \leq \varepsilon$, the CPU effort should be at least

$$CPU = K_2(\varepsilon/K_1)^{-d/\alpha}.$$

If the order α is one, the CPU increases like ε^{-3} in 3D (steady case) or even ε^{-4} in the unsteady 3D case. In many cases, this indicates that the computation at a useful accuracy level is simply not affordable. For a smooth function, the affordability increases importantly when order is increased.

In the case of a non-smooth function, the effort to represent it can be extremely large if this function involves a very large or infinite number of singularities. A more typical and interesting case is that of a function with a few singularities. To simplify our explanation, let us consider a Heavyside-type function, equal to 1 for an input larger than x_0, and equal to -1 otherwise. This function is very easy to represent/approximate with a few (four) real numbers, but very complex to approximate with a series of smooth approximations lying on uniform meshes. Conversely, it can be accurately represented by smooth functions lying on an adapted mesh.

The object of these volumes is to present a few analyses and methods devoted to the relation between an approximation and its mesh as well as how to adapt the mesh to both the approximation and the precise computational case.

Let us review what is presented in the chapters of this volume.

Recall that in Chapter 1 of Volume 1, we specified two particular approximation methods for compressible and two-fluid incompressible flows. We consider these particular approximation methods and try to cover a set of important questions to answer when we want to adapt a mesh to both the chosen approximation and the precise case to be computed.

The techniques presented in the remaining chapters of Volume 1 provided a new approach for an approximate solution, which is automatic through the mastering of mesh-convergence and safer through the mastering of approximation error, toward the certification of the approximate solution, presented in Chapter 1 of this volume.

In numerical simulation, we dream of an exact solution (let us call it "the grape") and we get a basket with half figs and half grapes, the fig here being "an error". It is then compulsory to know how many figs have replaced the grapes. Chapter 2 addresses the delicate problem of having a sufficiently clear quantification of the error. For this, we introduce the notion of corrector and describe means to compute a corrector.

Most physical processes of interest are unsteady and we have to build appropriate mesh adaptation strategies for unsteady phenomena (Chapter 3).

But adapted time steps can be small and penalizing for the global CPU efficiency. Chapter 4 addresses a multi-rate strategy (i.e. with non-uniform time steps) and explains its paramount impact on the efficiency of unsteady mesh adaptation.

An important objection to the minimization of interpolation error is the limited pertinence of this error in representing the *approximation error*, which is the difference between the exact solution of the PDE and the solution of the discretization of the PDE (approximate solution). A class of approximation-based mesh adaptation methods is the *goal-oriented formulation*, which minimizes the approximation error found on the evaluation of a scalar output depending on the PDE solution. In aeronautics, several outputs can be considered (drag, lift, moment, etc.). The goal-oriented method involves the expression of the error on the output in terms of the metric through *a priori* or *a posteriori* error estimates. We propose an analysis addressing this issue in Chapter 5 for steady inviscid flows.

An extension to viscous flows is described in Chapter 6.

Adapting mesh for the best evaluation of a single scalar output, as is the case for the goal-oriented method, is in many cases a severe limitation. The engineer is often interested in a small error in the whole representation of the flow. To answer this need, a second approximation error-based formulation is to search the mesh that minimizes a norm of the error committed on the flow field. This formulation is a *norm-oriented formulation*. It is presented in Chapter 7.

A goal-oriented approach can be also built for unsteady problems, and we present an example in Chapter 7. An extension to higher order is discussed in Chapter 8.

As a guide to the reader, an introduction to the basic methods can be obtained by reading Chapters 3–5 of Volume 1, which are restricted to feature-based/multiscale adaptation for steady models. The sequel for steady models should concern goal-oriented methods, and the reader can directly pass to Chapter 6 of this volume.

The reader interested in unsteady problems should continue in this volume with Chapter 3 for multiscale adaptation and Chapter 8 for goal-oriented adaptation.

Numerical experiments described in the book are performed with three different computational codes with different features. They are shortly described in Chapter 1 and mentioned in each numerical presentation in order to fix the ideas concerning the important details of their implementation.

1

Nonlinear Corrector for CFD

This chapter examines methods for improving an already obtained numerical solution with a reasonable computational effort. In the case of a Poisson problem, two types of approaches are identified. First, the corrector can rely on an a priori estimate. Second, the corrector can rely on a fictitious computation on a twice finer mesh through the defect correction (DC) process. A nonlinear version is the central point in the discussion. The description of nonlinear methods is focused on the Reynolds-averaged Navier–Stokes (RANS) equations. Examples of applications are RANS computations around airfoil and wing and the Euler evaluation of a supersonic flow around low-boom aircraft.

1.1. Introduction

The numerical simulation of engineering problems involves a discrete solution, which is alleged to converge toward the continuous solution of the problem as the size of the elements of the mesh decreases. As the exact solution of the continuous problem is sought, the difference between the numerical and the analytical solution is often seen as an unavoidable noise. But in an engineering context, this error can have disastrous consequences on the numerical prediction which can, in turn, lead to actual accidents or misconceptions (Collins et al. 1997; Jakobsen and Rosendahl 1994; Feghaly et al. 2008). It is thus essential not only to reduce but also to estimate this error. This is usually done by applying a mesh convergence study, see the previous chapter and Oberkampf and Trucano (2002), where the error is estimated with the difference between two solutions computed on two successive meshes. However, this approach is valid only if the asymptotic mesh-convergence has already been reached and requires an additional finer mesh (and thus expensive) calculation in order to appropriately estimate the error of the current solution. A mesh convergence study is thus mandatory to have a reasonable confidence into the discrete outputs, but it may fail. For instance, the AIAA CFD prediction workshops

(Rumsey et al. 2011; Levy et al. 2017; Rumsey and Slotnick 2015; Tinoco et al. 2017) have pointed out the dependency of the results on the type of meshes used, since two families of meshes may lead to two different answers. In consequence, very strict meshing guidelines based on previous experiences[1] are provided to minimize this effect, but this increases substantially the time required to generate meshes that meet these guidelines. Estimations of the error are based on the fact that the discrete solution W_h does not satisfy the continuous problem $\Psi(W) = 0$ or on the fact that a projection of the exact solution W on the discrete space does not satisfy the discrete problem $\Psi_h(W_h) = 0$ either. This can be stated in a priori and a posteriori ways:

$$\Psi_h(\Pi_h W) = \varepsilon_h \neq 0, \qquad [1.1]$$

$$\Psi(W_h) = \varepsilon \neq 0, \qquad [1.2]$$

where $\Pi_h W$ is a discrete projection of the continuous field W and W_h is the discrete solution. Notation Ψ holds for the differential operator defining the equation we are solving and Ψ_h is the discretization of Ψ. The idea behind correction is to use these defect equations [1.1] or [1.2] in order to estimate a corrector δW_h, which can be added to the first solution to get a corrected solution $\widetilde{W_h}$:

$$|\Psi(\widetilde{W_h})| = |\Psi(W_h + \delta W_h)| << |\varepsilon|, \qquad [1.3]$$

which should be sufficiently closer to the exact solution than W or to its projection $\Pi_h W$.

In the a priori case we write it formally as follows:

$$\Psi_h(\Pi_h W) \approx \Psi_h(W_h) + \frac{\partial \Psi_h}{\partial W}(\Pi_h W - W_h) \Rightarrow \delta W^h = [\frac{\partial \Psi_h}{\partial W}]^{-1} \Psi_h(\Pi_h W).$$

The difference $\Pi_h W - W_h$ can be approximated if we have a method to approximate $\Psi_h(\Pi_h W)$. It is explained in section 1.2.2.

In the a posteriori case, we write formally (considering that $W = W_h + \delta W_h$):

$$0 = \Psi(W_h + \delta W_h) \approx \Psi(W_h) + \frac{\partial \Psi}{\partial W}\delta W_h \Rightarrow \delta W_h = -[\frac{\partial \Psi}{\partial W}]^{-1} \Psi(W_h).$$

The evaluation of $\Psi(W_h)$ is a delicate task. An insufficiently accurate evaluation results in useless corrector δW_h. A robust option is to approximate the continuous

---

1 Meaning that somehow we already know the solution!

residual $\Psi(W_h)$ by a discrete residual $\Psi_{h/2}(W_h)$ computed on a two times finer mesh. The rest of this section focuses on the *a priori* approach with this second option.

1.1.1. *Linear correction*

Corrections have been addressed by different means in the past with the dual weighted residual (DWR) method (Becker and Rannacher 1996; Giles and Suli 2002a; Venditti and Darmofal 2002; Jones et al. 2006; Leicht and Hartmann 2010; Fidkowski and Darmofal 2011), and more generally the goal-oriented method (Loseille et al. 2010b) and the error transport equations (ETEs) (Layton et al. 2002; Pierce and Giles 2004; Hay and Visonneau 2006; Derlaga and Park 2017). DWR/goal-oriented methods focus on the correction of a functional. In some cases, the corrected functional is a higher order approximation of it. ETE methods focus on solution fields. Both rely on the same principles. As concerns DWR/goal-oriented methods, the linearization of the functional reads

$$j_h(\Pi_h W) \approx j_h(W_h) + \frac{\partial j_h}{\partial W} \cdot (\Pi_h W - W_h). \quad [1.4]$$

Since the projection $\Pi_h W$ is unknown, we introduce a finer mesh $h/2$. Put directly, we have

$$\Pi_h W - W_h \approx \Pi_h W_{h/2} - W_h, \quad \text{where} \quad \Psi_{h/2}(W_{h/2}) = 0.$$

In our standpoint, computing the fine-grid solution $W_{h/2}$ for evaluating a corrector is too costly. Conversely, knowing W_h, the evaluation of $\Psi_{h/2}(\Pi_{h/2} W_h)$ is affordable, and we have

$$\begin{aligned}\Psi_{h/2}(\Pi_{h/2} W_h) &\approx \Psi_{h/2}(W_{h/2}) + A_{h/2}(W_{h/2} - \Pi_{h/2} W_h) \\ &\Rightarrow \Pi_{h/2} W_h - W_{h/2} \approx \left(A_{h/2}\right)^{-1} \cdot \Psi_{h/2}(\Pi_{h/2} W_h),\end{aligned} \quad [1.5]$$

where $A_{h/2} = \dfrac{\partial \Psi_{h/2}}{\partial W_{h/2}}$ is the Jacobian of $\Psi_{h/2}$. From relations [1.4] and [1.5], we obtain

$$j_{h/2}(W_{h/2}) \approx j_{h/2}(\Pi_{h/2} W_h) + W_{h/2}^* \cdot \Psi_{h/2}(\Pi_{h/2} W_h),$$

where $W_{h/2}^* = A_{h/2}^{-T} \frac{\partial j_{h/2}}{\partial W_{h/2}}$ is the adjoint of the problem. In order to avoid the computation and the inversion of the large matrix $A_{h/2}$, the functional j and the numerical system are assumed to have a similar behavior on coarser and finer meshes so that $W_{h/2}^*$ is replaced by W_h, its analog on h:

$$j_{h/2}(W_{h/2}) \approx j_{h/2}(\Pi_{h/2}W_h) + \Pi_{h/2}W_h^* \cdot \Psi_{h/2}(\Pi_{h/2}W_h).$$

Then, the RHS is a corrected approximation of $j_h(W_h)$, which is obtained with the extra evaluation of the adjoint state W_h^* and the finer grid residual $\Psi_{h/2}(\Pi_{h/2}W_h)$.

Similarly, the ETE method computes the corrected solution field from the linearization of the problem as

$$\widetilde{W}_h \approx W_h - (A_h)^{-1}\,\Psi_h^*(W_h),$$

where Ψ_h^* is a defect residual obtained by different reconstruction means to estimate the error introduced by the discretization. The linearization of the problem is a determining factor for the quality of the corrected solution (Yan and Gooch 2015; Yan and Ollivier-Gooch 2017). This is true for both approaches. In particular, an approximate Jacobian obtained by passing to a lower spatial order of accuracy (as used in the Navier–Stokes solver introduced in Chapter 1) should not be used as it will lead to a poor correction. Lastly, would the computation of the matrix $\dfrac{\partial \Psi_h}{\partial W_h}$ even be exact, these two approaches are limited by the fact that the residual is linearized.

1.1.2. *Nonlinear correction*

Let us consider the ETE iteration with a RHS evaluated on a finer grid:

$$\widetilde{W}_h \approx W_h - (A_h)^{-1} RHS(W_h), \quad \text{with } RHS(W_h) = R_{h/2 \to h}\Psi_{h/2}(\Pi_{h/2}W_h),$$

where $R_{h/2 \to h}$ is a transfer from finer grid $h/2$ to grid h. If the updating $W_h \to \widetilde{W}_h$ can be reiterated and converged, we can gain the following advantages:

– the implementation is simple when we already have a Jacobian;

– it does not require a more accurate approximate Jacobian than the one of the flow solver;

– it automatically takes into account all numerical features and nonlinearities of the solver (limiter, flux reconstruction, Riemann solver, etc.).

This chapter is organized as follows. We first discuss in section 1.2 two ways of defining correctors for the usual P_1 approximation of a Poisson problem. We then recall in section 1.3 the notations related to the RANS equations and to their discretization. In section 1.4, the nonlinear discrete corrector for that scheme is explained. In section 1.5, we show the application of the finite volume nonlinear corrector to an inviscid supersonic flow simulation for a Sonic Boom prediction workshop.

1.2. Two correctors for the Poisson problem

1.2.1. *Notations*

Let $V = H_0^1(\Omega)$, Ω being a sufficiently smooth computational domain of $\mathbb{R}^2$. The continuous PDE system is written as

$$u \in V \mid Au = f \quad or \quad u \in V \mid \forall \varphi \in V, \ a(u,\varphi) = (f,\varphi), \qquad [1.6]$$

where

$$A = -div(\frac{1}{\rho}\nabla) \ ; \ a(u,\varphi) = \int_\Omega \frac{1}{\rho} \nabla u \cdot \nabla \varphi \, dx dy$$

and where $\frac{1}{\rho}$ is a positive, possibly discontinuous, scalar field in $L^\infty(\Omega)$. Further, we assume that the bilinear form a is coercive in space V, that is, there exists a positive α such that

$$a(v,v) \geq \alpha |v|_V^2.$$

This model exemplifies the pressure equation in some multi-phase incompressible flow formulation for which mesh adaptation is useful (e.g. Guégan et al. 2010). Let $\Omega_h = \Omega$ for simplicity, τ_h a triangulation of Ω_h and V_h be the usual P_1-continuous finite-element approximation space related to τ_h:

$$V_h = \{\varphi_h \in \mathcal{C}^0(\bar{\Omega}) \cap V, \varphi_h|_T \text{ is affine } \forall T \in \tau_h\}.$$

The finite-element discretization of [1.6] is written in variational and operational form:

$$u_h \in V_h \text{ and } \forall \varphi_h \in V_h, \ a(u_h,\varphi_h) = (f_h,\varphi_h), \qquad [1.7]$$

in such a way that u_h is a linear function of f_h, which we denote $u_h = A_h^{-1} f_h$. We denote by Π_h the usual interpolation operator:

$$\forall v \in \mathcal{C}^0(\bar{\Omega}) \cap V, \ \Pi_h v \in V_h \ \text{ and, } \ \forall \mathbf{x}_i \text{ vertex of } \Omega_h, \ \Pi_h v(\mathbf{x}_i) = v(\mathbf{x}_i).$$

Scalar correctors, that is, correctors for scalar outputs $j(u_h)$ depending on the solution, for example, $j(u_h) = (g, u_h)$ with g prescribed, have been defined by Giles and Pierce (1999). Our interest concerns the correction of the unknown field itself. Two options are now proposed.

1.2.2. A priori corrector for the PDE solution

A first rather simple a priori corrector can be derived from an analysis of the error RHS. We observe that:

$$a(u - u_h, \varphi_h) = (f - f_h, \varphi_h) \quad \forall \varphi_h \in V_h.$$

Assuming that the solution u is continuous, we can apply to u the usual P_1 interpolator Π_h and get:

$$a(\Pi_h u - u_h, \varphi_h) = a(\Pi_h u - u, \varphi_h) + (f - f_h, \varphi_h) \quad \forall \varphi_h \in V_h. \qquad [1.8]$$

We call $\Pi_h u - u_h$ the *implicit error*. By implicit, we mean that it can be obtained through solving a discrete system. It differs from the approximation error by an interpolation error:

$$u - u_h = \Pi_h u - u_h + u - \Pi_h u. \qquad [1.9]$$

In order to find an approximate of the implicit error, we need to evaluate the RHS of [1.8] for any test function φ_h. The second term of RHS of [1.8] is easy to evaluate (we know f and f_h). The first term of RHS of [1.8] can be transformed as follows (assuming that $\frac{1}{\rho}$ is constant on each element):

$$a(\Pi_h u - u, \varphi_h) = \sum_T \int_T \frac{1}{\rho} \nabla \varphi_h \cdot \nabla (\Pi_h u - u) \, dxdy$$

$$= \sum_T \int_{\partial T} (\Pi_h u - u) \frac{1}{\rho} \nabla \varphi_h \cdot \mathbf{n} \, d\sigma.$$

Then, we get

$$a(\Pi_h u - u_h, \varphi_h) = \sum_{\partial T_{ij}} \frac{1}{\rho} \nabla(\varphi_h|_{T_i} - \varphi_h|_{T_j}) \cdot \mathbf{n}_{ij} \int_{\partial T_{ij}} (\Pi_h u - u) \, d\sigma$$
$$+ (f - f_h, \varphi_h), \qquad [1.10]$$

where the sum is taken for all edges ∂T_{ij} separating triangles T_i and T_j of the triangulation. The unit vector $\mathbf{n}_{ij}$ normal to ∂T_{ij} is pointing outward T_i. Now, we do not know u but u_h. In order to evaluate the interpolation error, we first approximate the Hessian of u by an approximation $H_h(u_h)$ in V_h of the Hessian of u_h. We apply the recovery method of Zienkiewicz-Zhu type (1992) described in section 5.4.3 of Volume 1. This produces the values at vertices of the approximate Hessian $H_h(u_h)$. Then, $\Pi_h u - u$ is approximated by the continuous interpolation error $\pi_h u_h - u_h$ defined in section 5.4.6 of Volume 1. This P_2 by element function is zero at vertices and $\frac{1}{8}(H_h(u_h)(\mathbf{x}_i) + H_h(u_h)(\mathbf{x}_j)).\mathbf{x}_i\mathbf{x}_j.\mathbf{x}_i\mathbf{x}_j$ at mid-edges ij. We apply the same type of estimate to $f - f_h$, namely, $f - f_h \approx -(\pi_h f_h - f_h)$. In order to approximate the implicit error $\Pi_h u - u_h$, solution of [1.10], we define our *a priori implicit corrector* as follows:

$$\bar{u}'_{prio} \in V_h, \text{ and } \forall \varphi_h \in V_h,$$
$$a(\bar{u}'_{prio}, \varphi_h) = \sum_{\partial T_{ij}} (\frac{1}{\rho} \nabla \varphi_h|_{T_i} - \nabla \varphi_h|_{T_j}) \cdot \mathbf{n}_{ij} \int_{\partial T_{ij}} (\pi_h u_h - u_h) \, d\sigma$$
$$- (\varphi_h, \pi_h f_h - f_h). \qquad [1.11]$$

This corrector is an approximation of $\Pi_h u - u_h$. Replacing again the unknown interpolation error by its evaluation on the discrete approximation, we define our *a priori corrector* as follows:

$$u'_{prio} = \bar{u}'_{prio} - (\pi_h u_h - u_h). \qquad [1.12]$$

Assuming that the approximations made between [1.10] and [1.11] are small, the *a priori corrector* u'_{prio} is then built in such a way that (formally)

$$u'_{prio} \approx u - u_h.$$

This corrector is easy to compute, but we shall see in Chapter 6 of this volume that its accuracy is rather low.

1.2.3. *Finer-grid DC corrector for the PDE solution*

A second option relies on using a fictitious finer grid. Here, we write it in the linear case. Let us assume that the approximation is in its asymptotic mesh convergence phase for the mesh Ω_h under study of size h. Then, this will be also true for a strictly two times finer embedding mesh $\Omega_{h/2}$ and, for our second-order accurate scheme applied to a sufficiently smooth problem, the *heuristics* of Richardson analysis is written as

$$u_h = A_h^{-1} f_h \;,\; u_{h/2} = A_{h/2}^{-1} f_{h/2} \;\Rightarrow\; u - u_{h/2} \approx \frac{1}{4}(u - u_h), \qquad [1.13]$$

where u_h and $u_{h/2}$ are, respectively, the solutions on Ω_h and $\Omega_{h/2}$. Note that in the case of local singularities, second-order convergence [1.13] is not true for uniform meshes since global convergence degrades to first-order convergence. But Chapter 2 of Volume 1 tends to show that it holds almost everywhere for a sequence of adapted meshes (see also Loseille et al. 2007).

Then, at least formally, we have

$$\Pi_h u - \Pi_h u_{h/2} \approx \frac{1}{4}(\Pi_h u - u_h),$$

which implies

$$\Pi_h u - u_h \approx \frac{4}{3}(\Pi_h u_{h/2} - u_h).$$

Now,

$$A_{h/2}(u_{h/2} - P_{h \to h/2} u_h) \;=\; f_{h/2} - A_{h/2} P_{h \to h/2} u_h,$$

where $P_{h \to h/2}$ linearly interpolates coarse values on fine mesh. This allows to evaluate $\Pi_h u - u_h$ but needs to solve a finer grid system. Let us introduce the residual transfer $R_{h/2 \to h}$, which accumulates on coarser grid vertices the values at finer vertices in neighboring coarse elements multiplied with barycentric weights. In order to reduce the computational cost to solving a coarser grid system, we approximate $\Pi_h A_{h/2}^{-1}$ by $A_h^{-1} R_{h/2 \to h}$:

$$\Pi_h(u_{h/2} - P_{h \to h/2} u_h) \;\approx\; A_h^{-1} R_{h/2 \to h} \left(f_{h/2} - A_{h/2} P_{h \to h/2} u_h \right).$$

This motivates the definition of a finer grid DC corrector as follows:

$$A_h \bar{u}'_{DC} = \frac{4}{3} R_{h/2 \to h}(f_{h/2} - A_{h/2} P_{h \to h/2} u_h). \qquad [1.14]$$

The *DC corrector* $\bar{u}'_{DC}$ approximates $\Pi_h u - u_h$ instead of $u - u_h$ and can be corrected as the previous one:

$$u'_{DC} = \bar{u}'_{DC} - (\pi_h u_h - u_h). \qquad [1.15]$$

1.3. RANS equations

1.3.1. *Vector form of the RANS system*

The compressible Navier–Stokes equations for mass, momentum and energy conservation with Spalart–Allmaras turbulence model are those described in Chapter 1 of Volume 1. It is useful to re-write the RANS system in the following more compact vector form:

$$\Psi(W) = W_t + F_1(W)_x + F_2(W)_y + F_3(W)_z - S_1(W)_x - S_2(W)_y$$
$$- S_3(W)_z - Q(W) = 0$$

(+ boundary conditions), where $F_i(W)_a = \dfrac{\partial F_i(W)}{\partial a}$ ($i = 1, 2, 3, a = x, y, z$) (idem for S). W is the non-dimensionalized conservative variables vector:

$$W = (\rho, \rho u, \rho v, \rho w, \rho E, \rho \tilde{\nu})^T.$$

$F(W) = (F_1(W), F_2(W), F_3(W))$ are the convective (Euler) flux functions:

$$\begin{aligned}
F_1(W) &= \left(\rho u,\ \rho u^2 + p,\ \rho uv,\ \rho uw,\ u(\rho E + p),\ \rho u \tilde{\nu}\right)^T, \\
F_2(W) &= \left(\rho v,\ \rho uv,\ \rho v^2 + p,\ \rho vw,\ v(\rho E + p),\ \rho v \tilde{\nu}\right)^T, \qquad [1.16] \\
F_3(W) &= \left(\rho w,\ \rho uw,\ \rho vw,\ \rho w^2 + p,\ w(\rho E + p),\ \rho w \tilde{\nu}\right)^T.
\end{aligned}$$

$S(W) = (S_1(W), S_2(W), S_3(W))$ are the viscous fluxes:

$$S_1(W) = \left(0, \mathcal{T}_{xx}, \mathcal{T}_{xy}, \mathcal{T}_{xz}, u\mathcal{T}_{xx} + v\mathcal{T}_{xy} + w\mathcal{T}_{xz} + \lambda T_x, \frac{\rho}{\sigma}(\nu + \tilde{\nu})\tilde{\nu}_x\right)^T,$$

$$S_2(W) = \left(0, \mathcal{T}_{xy}, \mathcal{T}_{yy}, \mathcal{T}_{yz}, u\mathcal{T}_{xy} + v\mathcal{T}_{yy} + w\mathcal{T}_{yz} + \lambda T_y, \frac{\rho}{\sigma}(\nu + \tilde{\nu})\tilde{\nu}_y\right)^T, \quad [1.17]$$

$$S_3(W) = \left(0, \mathcal{T}_{xz}, \mathcal{T}_{yz}, \mathcal{T}_{zz}, u\mathcal{T}_{xz} + v\mathcal{T}_{yz} + w\mathcal{T}_{zz} + \lambda T_z, \frac{\rho}{\sigma}(\nu + \tilde{\nu})\tilde{\nu}_z\right)^T,$$

where $\mathcal{T}_{ij}$ are the components of laminar stress tensor $\mathcal{T} = (\mu + \mu_t)[(\nabla \otimes \mathbf{u} + {}^t\nabla \otimes \mathbf{u}) - \frac{2}{3}\nabla.\mathbf{u}\,\mathbb{I}]$:

$$\mathcal{T}_{xx} = (\mu + \mu_t)\frac{2}{3}(2u_x - v_y - w_z), \quad \mathcal{T}_{xy} = (\mu + \mu_t)(u_y + v_x),$$

$$\mathcal{T}_{xz} = (\mu + \mu_t)(u_z + w_x), \quad ...$$

$Q(W)$ are the source terms, that is, the diffusion, production and destruction terms from the Spalart–Allmaras turbulence model:

$$Q(W) = \left(0, 0, 0, 0, 0, \frac{c_{b2}\rho}{\sigma}\|\nabla\tilde{\nu}\|^2 + \rho c_{b1}\tilde{S}\tilde{\nu} + c_{w1}f_w\rho\left(\frac{\tilde{\nu}}{d}\right)^2\right)^T. \quad [1.18]$$

Note that $Q = 0$ in the case of the laminar Navier–Stokes equations, additional source terms are added (to take into account gravity, for instance).

1.3.2. *Formal discretization*

1.3.2.1. *Finite element discretization*

In the standard Galerkin (possibly stabilized) approach, the strong form of the equations over a given domain Ω is reshaped in a weak form:

$$\Psi(W) = 0 \iff \int_\Omega \Psi(W)\varphi d\Omega = 0, \quad \forall \varphi \in V(\Omega),$$

with additional boundary conditions and where $V(\Omega) = H_0^1(\Omega) = \{\varphi \in L_0^2(\Omega) \mid \nabla\varphi \in L^2(\Omega)\}$ is the functional space in which the solution is sought. The problem is discretized by restricting the functional space to a subset $V_h(\Omega)$ of $V(\Omega)$. In the standard finite element approach, the domain Ω is split in a finite number of discrete elements K_i (triangles, tetrahedra, hexahedra, prisms, etc.) on which the restriction $\varphi_{h|K_i}$ is a polynomial for each function $\varphi_h \in V_h(\Omega)$ and so that $\cup_i K_i = \Omega$. The value of $\varphi_{h|K_i}$ on each element is determined by a set of given nodal values and the functions φ_i – defined by setting the ith node to 1 and the other to 0 – form a basis of $V_h(\Omega)$. The residual vector of the discrete problem is defined by

$$\Psi_h^i(W_h) = \int_\Omega \Psi(W_h)\varphi_i \mathrm{d}\Omega.$$

1.3.2.2. *Finite volume discretization*

In the standard finite volume approach, the domain Ω is split in a finite number of cells $\cup_i C_i = \Omega$ and the residual is defined as

$$\Psi_h^i(W) = \int_{C_i} \Psi(W)\mathrm{d}\Omega,$$

for any continuous field W, where C^i is the cell associated with vertex P_i. The problem is discretized defining the vector $W_i = \int_{C_i} W \mathrm{d}\Omega$, supposed to be the mean value of the field on each cell, and a reconstruction operator $\mathcal{U}$ that recovers a field W on each cell from the discrete values W_i. For flow problems, the advection terms can be written as a divergence; thus using the Green formulae, the integral of advection term is the integral of the advection fluxes over the boundaries of the cell:

$$\int_{C^i} \nabla \cdot \Psi(\mathcal{U}W_h)\mathrm{d}\Omega = \int_{\partial C^i} \Psi(\mathcal{U}W_h) \cdot \mathbf{n}\mathrm{d}\gamma,$$

where $\mathbf{n}$ is the inward normal and $\mathcal{U}$ a reconstruction operator that recovers a field W on each cell from the discrete values W_h.

1.3.3. *Notations for discretization*

The spatial discretization of the RANS model is described in Chapter 1 of Volume 1. We give here some details of the implementation of our in-house finite volume–finite element flow solver `Wolf` as this induces several constraints – as we will see – on the implementation of the corrector. In the sequel, we denote for a given vertex P_i:

– $\mathcal{E}(P_i)$ the set of edges joining P_i to another vertex P_j of the mesh;

– $\mathcal{V}^1(P_i)$ the set of vertices that are neighbors of P_i, that is, the set of vertices connected to P_i by an edge. $\mathcal{V}^1(P_i)$ is named the first-order ball of P_i.

– $\mathcal{V}^2(P_i)$ the set of vertices that are neighbors of the neighbors of P_i. $\mathcal{V}^2(P_i)$ is named the second-order ball of P_i. Similarly, $\mathcal{V}^k(P_i)$ is the kth iteration of this process.

The discretization of Euler fluxes are sufficiently described in Chapter 1 of Volume 1. Concerning the viscous terms using the P_1 finite element method (FEM), we use the following notations:

$$\mathbf{S}_i = \sum_{P_j \in \mathcal{V}^1(P_i)} \int_{\partial C_{ij}} S(W_i) \cdot \mathbf{n}\, d\gamma,$$

where ∂C_{ij} is the common interface between cells C_i and C_j. Let φ_i be the P_1 finite element basis function associated with vertex P_i, we have

$$\int_K \nabla \varphi_i\, d\Omega = -\int_{\partial C_i \cap K} \mathbf{n}\, d\gamma,$$

and as the solution is represented on the P_1 basis, $S(W_i)$ (which comes from a gradient) is assumed constant by parts on each element K, for example,

$$\mathcal{T}_{xy}|_K = \sum_{P_i \in K} \mu|_K \left(u_i \frac{\partial \varphi_i}{\partial y} + v_i \frac{\partial \varphi_i}{\partial x} \right),$$

where $\mu|_K$ is the mean value of μ on the element to be constant. Then, we obtain

$$\sum_{P_j \in \mathcal{V}^1(P_i)} \int_{\partial C_{ij}} S(W_i) \cdot \mathbf{n}\, d\gamma = \sum_{K \ni P_i} S(W_i)|_K \cdot \int_{\partial C_i \cap K} \mathbf{n}\, d\gamma$$

$$= -\sum_{K \ni P_i} \int_K S(W_i)|_K \cdot \nabla \varphi_i\, d\Omega.$$

The effective computation of the previous integral then leads to the computation of integrals of the following form:

$$\int_K \nabla \varphi_i\, \nabla \varphi_j\, d\Omega = |K|\, \nabla \varphi_i|_K\, \nabla \varphi_j|_K.$$

In this expression, $\nabla \varphi_i|_K$ is the constant gradient of basis function φ_i associated with vertex P_i. In the facts, these terms are computed element-wise and each contribution is added to each vertex of the element. Given an element $K = (P_i, P_j, P_k, P_l)$, we have the partial flux associated with each vertex:

$$\Phi_{i,K}^{visc}(W_i, W_j, W_k, W_l) = \int_{\partial C_i \cap K} S|_K \cdot \mathbf{n}\, d\gamma.$$

1.4. Nonlinear functional correction

1.4.1. *Finite volume nonlinear corrector*

The central idea is to approximate $\Psi(W_h)$ by $\Psi_{h/n}(W_h)$. To accumulate on the h-mesh the residual computed on the h/n-mesh, we split the accumulation process depending on the possible configurations. If we consider a h-mesh vertex P_i, for the accumulation of the residual at P_i the different cases are as follows:

– the h/n-mesh vertex Q_k coincides with the h-mesh vertex P_i, then the full residual is accumulated: $\varepsilon_{h/n}(Q_k) = \varepsilon_{h/n}(P_i)$;

– the h/n-mesh vertex Q_k belongs to a h-mesh edge e_j containing P_i, then we compute the barycentric coordinates $\beta_i(e_j)$ of Q_k w.r.t. P_i on edge e_j (Alauzet 2016). The accumulated residual is $\beta_i(e_j)\varepsilon_{h/n}(Q_k)$;

– the h/n-mesh vertex Q_k belongs to a h-mesh face f_j containing P_i, then we compute the barycentric coordinates $\beta_i(f_j)$ of Q_k with respect to P_i on face f_j. The accumulated residual is $\beta_i(f_j)\varepsilon_{h/n}(Q_k)$;

– the h/n-mesh vertex Q_k is inside a h-mesh element K_j containing P_i, then we compute the barycentric coordinates $\beta_i(K_j)$ of Q_k with respect to P_i in element K_j. The accumulated residual is $\beta_i(K_j)\varepsilon_{h/n}(Q_k)$.

For the h-mesh vertex P_i, the accumulation process is expressed as

$$[\mathcal{A}_h^{h/n}(\varepsilon_{h/n})][P_i] = \varepsilon_{h/n}(P_i) + \sum_{Q_k \in e_j | e_j \ni P_i} \beta_i(e_j)\varepsilon_{h/n}(Q_k)$$
$$+ \sum_{Q_k \in f_j | f_j \ni P_i} \beta_i(f_j)\varepsilon_{h/n}(Q_k) + \sum_{Q_k \in K_j | K_j \ni P_i} \beta_i(K_j)\varepsilon_{h/n}(Q_k). \quad [1.19]$$

For $n = 2$, this gives Algorithm 1.1. A 2D example of this accumulation process at the element level is illustrated in Figure 1.1.

Applying the NS solver to compute (1) the initial solution, (2) the residual on the $h/2$-mesh and (3) the corrected solution ensure that the corrector encompasses all the features of the solver (extrapolation, Riemann solver, limiters etc.). The number of iterations required to compute the corrected solution depends on the considered problem (essentially its nonlinearity), the simplified Jacobian and the linear solver used (preconditioner and fixed point iteration) in case of incomplete linear convergence. As the initial and the corrected solution are relatively close to each other, the number of iterations required to compute the corrected solution is typically about $1/10$ to $1/100$ of the number of iterations required to compute the initial solution. Finer subdivision ($h/4$, $h/8$, etc.) of the initial mesh can also be used to improve the predictions. It is very important to note that linearly interpolating the discrete solution W_h on the $h/2$-mesh (i.e. using $\Pi_{h/2}(W_h)$) is nonetheless convenient but also essential to the precision of the correction. Replacing $\Pi_{h/2}(W_h)$ by a better polynomial reconstruction tends to reduce the efficiency of the correction. Indeed, as we want to quantify the error of piecewise linear solution W_h with respect to the $h/2$-mesh; it is counter productive to try to improve $\Pi_{h/2}(W_h)$. *More numerical implementation details are given in Frazza et al. (2019).*

Algorithm 1.1. Finite volume nonlinear corrector computation

1) Compute the solution of $\Psi_h(W_h) = 0$ on the initial mesh: W_h.

2) Linearly interpolate the solution on the $h/2$-mesh: $\Pi_{h/2} W_h$.

3) Compute the $h/2$-mesh residual: $\Psi_{h/2}(\Pi_{h/2} W_h) = \varepsilon_{h/2}$.

4) Accumulate the $h/2$-mesh residual on the h-mesh: $S_h = \mathcal{A}_h^{h/2}(\varepsilon_{h/2})$.

5) Perform a few iterations of the flow solver with S_h as source term to compute the corrected solution: $\Psi_h(\widetilde{W_h}) = S_h$.

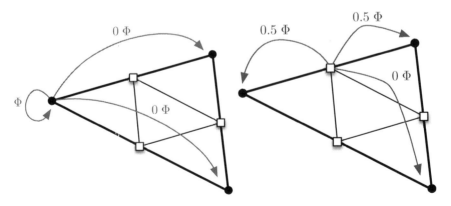

Figure 1.1. *Examples of linear accumulation process when the $h/2$-mesh vertex Q_k coincides with the h-mesh vertex P_i (left) and when the $h/2$-mesh vertex Q_k is the mid-point of a h-mesh edge e (right). For a color version of this figure, see www.iste.co.uk/dervieux/meshadaptation2*

1.4.2. *Finite element corrector*

We transpose the variational derivation of section 1.2.2:

$$\int_\Omega \Psi(W)\varphi \, d\Omega = 0, \quad \forall \varphi \in V,$$

$$\int_\Omega \Psi(W_h)\varphi_h \, d\Omega = 0, \quad \forall \varphi_h \in V_h \subset V.$$

In particular, the exact solution verifies

$$\int_\Omega \Psi(W)\varphi_h \, d\Omega = 0, \quad \forall \varphi_h \in V_h,$$

but generally

$$S_h = \int_\Omega \Psi(\Pi_h W)\varphi_h \, d\Omega \neq 0, \quad \forall \varphi_h \in V_h.$$

We have

$$\int_\Omega \Psi(W)\varphi_h \, d\Omega = \int_\Omega \Psi(\Pi_h W + (W - \Pi_h W))\varphi_h \, d\Omega = 0, \quad \forall \varphi_h \in V_h,$$

and by linearizing the fluxes in $\Pi_h W$ and W_h, we have

$$\Psi(\Pi_h W + (W - \Pi_h W)) \approx \Psi(\Pi_h W) + \left.\frac{\partial \Psi}{\partial W}\right|_{\Pi_h W} \cdot (W - \Pi_h W),$$

and

$$\Psi(W_h + (W - \Pi_h W)) \approx \Psi(W_h) + \left.\frac{\partial \Psi}{\partial W}\right|_{W_h} \cdot (W - \Pi_h W) \approx \Psi(W_h)$$
$$+ \left.\frac{\partial \Psi}{\partial W}\right|_{\Pi_h W} \cdot (W - \Pi_h W).$$

From these three relations above, we deduce for all $\varphi_h \in V_h$:

$$\int_\Omega \Psi(\Pi_h W)\varphi_h \, d\Omega \approx \int_\Omega \Psi(\Pi_h W + (W - \Pi_h W))\varphi_h \, d\Omega - \int_\Omega \left.\frac{\partial \Psi}{\partial W}\right|_{\Pi_h W}$$
$$\cdot (W - \Pi_h W)\varphi_h \, d\Omega$$
$$\approx \int_\Omega \Psi(W_h)\varphi_h \, d\Omega - \int_\Omega \Psi(W_h + W - \Pi_h W)\varphi_h \, d\Omega \quad [1.20]$$
$$\approx -\int_\Omega \Psi(W_h + W - \Pi_h W)\varphi_h \, d\Omega. \quad [1.21]$$

This expression relies on the fact that W_h is relatively close to $\Pi_h W$, so that the derivatives $\left.\frac{\partial \Psi}{\partial W}\right|_{\Pi_h W} \approx \left.\frac{\partial \Psi}{\partial W}\right|_{W_h}$ are approximatively the same. It has two major advantages. First, it does not require to compute the derivative $\left.\frac{\partial \Psi}{\partial W}\right|_{W_h}$, which saves a lot of trouble: as we mentioned previously, in linear approaches, the quality of the correction depends on the choice and the precision of these derivatives. Second, it automatically takes into account fairly well the nonlinearities of the residual. As long as the higher derivatives $\left.\frac{\partial^k \Psi}{\partial W^k}\right|_{\Pi_h W} \approx \left.\frac{\partial^k \Psi}{\partial W^k}\right|_{W_h}$ are relatively close, which is a reasonable assumption as the correction $W - \Pi_h W$ is a purely local correction, the Taylor developments mentioned previously can be expanded.

In relation [1.21], we still need to estimate $W - \Pi_h W$ that represents a loss of smoothness due to the discretization of the solution; it does not depend on the absolute value of the solution but on its variations. The term $W_h + W - \Pi_h W$ can be thus estimated by reconstructing a smooth solution $\mathcal{U}_h^{k+1}(W_h)$ (Clément 1975; Zienkiewicz and Zhu 1992; Bank and Smith 1993) from the actual numerical solution W_h as shown in Figure 1.2. Finally, the source term can be expressed as

$$\int_\Omega \Psi(\mathcal{U}_h^{k+1}(W_h))\varphi_h \, d\Omega = S_h(\varphi_h), \quad \forall \varphi_h \in V_h, \quad [1.22]$$

where $\mathcal{U}^{k+1}$ holds for the $(k+1)^{th}$ order reconstruction if the solution is of order k. We propose Algorithm 1.2. to compute the corrector. Note that there is no need for a finer mesh in this case as the integrals are computed on the same elements, but more quadrature points are required to have a consistent computation of the source term given by relation [1.22].

Algorithm 1.2. Finite element nonlinear corrector computation

1) Compute the solution on the initial mesh: W_h.
2) Reconstruct the smooth solution on each element: $\mathcal{U}_h^{k+1}(W_h)$.
3) Compute the residual: $S_h^i = R_h^i(\mathcal{U}_h^{k+1}(W_h)) = \int_\Omega \Psi(\mathcal{U}_h^{k+1}(W_h))\varphi_h^i \, d\Omega$.
4) Perform a few iterations of the flow solver with S_h as source term to compute the corrected solution: $\Psi_h(\widetilde{W_h}) = S_h$.

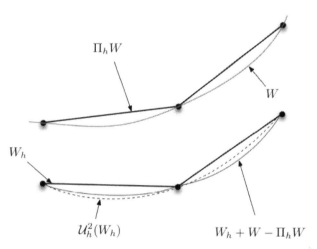

Figure 1.2. *Reconstruction of a smoother solution (dashed curve) from a discrete solution (straight segments) compared to the exact solution (W) and the reconstructed solution with the exact defect ($W_h + W - \Pi_h W$). For a color version of this figure, see www.iste.co.uk/dervieux/meshadaptation2*

REMARK 1.1.– For both nonlinear corrector methods, we take great care to propose methods that do not require any additional memory costs. With regard to the computational cost, the computation of the source term S_h is approximatively the cost of one flow solver iteration, which is negligible.

1.5. Example: supersonic flow

In Frazza et al. (2019), the interested reader will find, among others, numerical experiments with manufactured solutions. We shall stress here the interest of the corrector by giving a short view of a computation described in details in Frazza et al. (2019). The considered flow (C25D Euler flow) is one among those proposed for the

2nd AIAA Sonic Boom prediction workshop (Park and Nemec 2017). An idea of the geometry and the resulting flow is given in Figure 1.3. Figure 1.4 (left) shows the pressure signatures (normalized pressure difference with the free stream pressure) under the aircraft at a distance[2] $R/L = 1$, computed on the tailored meshes provided for the workshop. Error bars represent the estimated error of each nodal value computed with the nonlinear corrector. We can see that the corrector indicates that the error level remains relatively high, even on the 13 million vertices mesh. This has been confirmed by the workshop results: the solution mesh convergence is not achieved on the provided series of tailored meshes. This is visible on the leading shock, which is still relatively smooth, and the shocks in the aft part of the signature ($65 < X < 80$), which are still forming. Note that the actual subdivision by two of the size of this mesh with 13 million vertices would have yielded a mesh with 104 millions vertices! Figure 1.4 (right) shows the same computations using the goal-oriented mesh adaptation of Chapter 4 of this volume, where the goal is to optimally capture the integral of pressure signature. We can see that the error bars are much smaller. This is especially visible on the mid-part of the signature ($50 < X < 60$). It is even more enlightening to compare the error levels predicted by the corrector on the tailored workshop mesh composed of 13 million vertices in the mid-part of the signature to the actual difference between the solutions on the tailored mesh and adapted mesh composed of 5.9 million vertices. The error predicted by the corrector on tailored mesh matches the difference between the solution on the tailored mesh (which is not converged) and the solution on the adapted mesh (which is converged in this region).

The results obtained by the corrector (and presented in more detail in Loseille et al. 2018) are interesting because they are, in particular, pointing out a large error in the aft-part of the pressure signature ($65 < X < 80$) on the workshop tailored meshes, while this error does not appear with adequately adapted meshes.

1.6. Concluding remarks

This chapter presents nonlinear correctors for the finite volume and the finite element methods, which are based either on a local subdivision of the mesh, or a local reconstruction of the solution, and an extra-resolution of the problem with an added source term. The main advantages of this strategy are as follows:

– we need not solve the problem on the global subdivided mesh or at a higher order that yields large memory overhead and is unsuitable when dealing with realistic cases. The proposed methods do not require any additional memory costs;

– the local subdivision makes it easy to parallelize the computation of the source term;

---

2 R is the distance to the aircraft and L is the length of the aircraft.

– the cost of computing the nonlinear corrector is usually less than 10% of the overall simulation cost;

– the corrector automatically takes into account the nonlinearities of the problem at hand because we perform many iterations of the flow solver with the added source term to compute the corrected solution;

– the corrected solution takes into account all the features of the considered numerical method.

Figure 1.3. *Second AIAA Sonic Boom prediction workshop. Left: C25D geometry. Right: Cut plane in the volume behind the aircraft where we can observe in the left part the Mach field and in the right part the 5.9 million vertices adapted mesh obtained with a goal-oriented mesh adaptation on the pressure signature. For a color version of this figure, see www.iste.co.uk/dervieux/meshadaptation2*

This nonlinear approach has proven to bring a significant improvement for nonlinear problems, removing the dependency on the choices in the linearization. Such a corrector can be useful in the following processes.

First, a corrector is a good postprocessing for increasing the accuracy of a computation on a given mesh, without changing the mesh and without an important extra computational effort. All the numerical examples have demonstrated that this nonlinear corrector provides an efficient and pertinent correction of the solution and in turn an estimation of the numerical error. Further, the corrected solution remains second-order accurate and, in case of mesh convergence, can be used for a Richardson extrapolation.

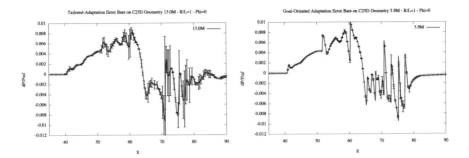

Figure 1.4. *Second AIAA Sonic Boom prediction workshop. Pressure difference under the plane (continuous line) for a tailored mesh (left) composed of 13 million vertices and adapted meshes (right) composed of 5.9 million vertices. The estimated error provided by the nonlinear corrector is shown by the vertical error bars. For a color version of this figure, see www.iste.co.uk/dervieux/meshadaptation2*

Second, the corrector provides an estimation of the numerical error level that can be used to assess the solution mesh convergence (indicating when to stop the convergence analysis) and to give error bars (quantification of the uncertainties due to the discretization) on the obtained numerical solution. The sonic boom results demonstrate how the corrector can be a very useful complement to mesh adaptation, since it can be checked, very coherently, that the mesh adaptation produce a flow for which the corrector is much smaller than those obtained with insufficient mesh resolution.

Third, the corrector can be used as a representation of the approximation error and replace it in the functional error of a mesh adaption process. This is the principle of the norm-oriented mesh adaptation method that is described in Chapter 6 of this volume.

1.7. Notes

More computations can be found in Frazza et al. (2019).

An important application of the computation of correctors is the derivation of norm-oriented mesh adaptation, as presented in Chapter 6 of this volume.

2

Multi-scale Adaptation for Unsteady Flows

In this chapter, we describe an extension of the multiscale feature-based mesh adaptation method for the calculation of an unsteady flow. The mesh adaptation relies on a metric-based method controlling the $\mathbf{L}^p$-norm of the spatial interpolation error. Mesh-adaptive time advancing is achieved because of a *transient fixed-point adaptation algorithm* to predict the solution evolution coupled with a metric intersection in time procedure. In time direction, we impose equidistribution of error, that is, minimization in $\mathbf{L}^\infty$. This adaptive approach is illustrated with an incompressible two-phase flow using a level set formulation.

2.1. Introduction

Multi-scale metric-based mesh adaptation methods need to be carefully extended before they efficiently apply to unsteady flows. We restrict our discussion to numerical unsteady schemes using *time advancing*. We consider adaptation methods changing the spatial mesh during the simulation time frame. Two options are usually considered. In the first option, mesh changes during the time advancing, for instance, between t^n and t^{n+1}. Due to the difficulties in building a time derivative between meshes of different connectivities, the mesh is generally only deformed, for example, as in the moving finite element (Baines 1994). In the second option, the mesh does not change during the time advancing, but instead at a time level, for example, t^n. Once the mesh is changed, the solution is advanced from t^n to t^{n+1} with the new mesh. It may happen that a good mesh for level t^n is a bad one for t^{n+1}. For example, this happens when a shock is moving. For a robust accuracy, the mesh need to be adapted not only for level t^n but also for level t^{n+1}. In many papers, a device is found in order to *anticipate* the evolution during the corresponding time interval $[t^n, t^{n+1}]$. In contrast, in the present chapter, the adequacy of the mesh with the concerned time levels is obtained by a fixed point algorithm, which generalizes to time advancing the Hessian-based steady-state fixed point described by

Algorithm 5.1. in Chapter 5 of Volume 1. This extension is the *transient fixed-point* (TFP) *mesh adaptation algorithm* initiated in Alauzet et al. (2007). In this algorithm, the solution evolution through a specified time interval, involving one or several solver time steps, is first *predicted*, then the mesh is adapted to the predicted solution evolution because of the metric intersection between the time levels involved in the considered time interval. Finally, the solution is advanced with the new adapted mesh.

In this chapter, the TFP is introduced and analyzed in order to approach the early capturing (EC) property for spatial error and, at least, to show a predictable mesh convergence rate that will be better than with non-adaptive approaches. To demonstrate the efficiency of the proposed approach, we concentrate on an unsteady flow model involving discontinuities, namely the incompressible two-phase model defined and discretized in Chapter 1 of Volume 1. Calculations involve the advection of a discontinuous density and deals with discontinuities (density, viscosity) in momentum equations. If in such a calculation the interface is not fitted by the mesh (like, for example, in a Lagrangian mode), then its numerical capturing may induce a deterioration of convergence order. The level set method is a mean for increasing the smoothness of the interface advection problem. An analysis in Chapter 1 of Volume 1 predicts an order of $4/3$ for Euler flows, but this analysis does not extend to Navier–Stokes flows. In the velocity/momentum equations, the flow variables have discontinuous properties (density, viscosity) near the interface and accuracy order is limited to one. These discontinuities could be addressed with a sophisticated discontinuous approximation like the ghost fluid method (GFM) (Fedkiw et al. 1999), with the hope that most part of the singularities are taken into account in order to recover a high-order accuracy. We do not consider GFM in this chapter; instead, a thickened interface method is applied. These discontinuous properties are smoothed by replacing the discontinuity by a sinusoidal transition. But if no mesh adaptation is used, this may limit the accuracy to first order.

In this chapter, we keep the basic option of a feature-based adaptation, that is, we make the assumption that the discretization error can be controlled by the interpolation error of some features prescribed by the user, as done in the previous chapters. We combine several error criteria for adapting the mesh: first, to the momentum equations, and second, to the level set advection. We describe a method for controlling, time interval after time interval, the error of transient flows. It relies on the maximum in time of the instantaneous $\mathbf{L}^p$ norm of spatial error. We explain how to introduce this strategy in the transient fixed point mesh adaptation algorithm. Conditions for faster convergence to the continuous solution are discussed. This method is applied to several simplified and more complex test cases in order to evaluate the resulting gain in accuracy and convergence. Sections 2.2 and 2.3 present the central issues of mesh adaptation with notably the transient fixed point mesh adaptation algorithm and the specific adaptation to the interface. Then, in section 2.4, the impact of the proposed algorithm on the solution accuracy and the numerical convergence is analyzed on a two-dimensional example. Finally, in section 2.5, this approach is validated on a realistic three-dimensional example for which experimental data are available.

2.2. Mesh adaptation efficiency

Two new difficulties are addressed in this chapter. First, we consider an unsteady flow. Our numerical model is a time-advancing model and we have to consider how frequently the spatial mesh needs to be changed. Second, we consider a flow with a discontinuous (moving) interface, which will need to be captured accurately by the adapted mesh.

2.2.1. *Regular and singular unsteady model*

As explained in section 2.9 of Volume 1, several new issues must be considered for mesh adaptation in the unsteady case.

2.2.1.1. *Space-time convergence analysis for unsteady calculations*

The numerical approximation order can be evaluated on the basis of a space–time convergence, over the cylinder $Q = \Omega \times [0, T]$. For this, the extended definition of convergence order $||u - u_N||_Q = O(N^{\frac{\alpha}{d}})$ still applies. For a 3D unsteady simulation, the dimension d is set to 4; the total number of degrees of freedom N representing the approximate space–time function is the total number, time level after time level, of the spatial nodes used during the unsteady calculation. In the case of uniform meshes with m nodes and n time levels, $N = m \times n$. Another equivalent way to express a convergence order at least equal to α is to compare two embedded discretizations, namely for N and $k^d N$ degrees of freedom:

$$||u - u_{k^d N}|| \leq k^{-\alpha} ||u - u_N||. \qquad [2.1]$$

Let us focus, for example, on the motion of a step function in a fixed 3D spatial domain *uniformly meshed*. We assume that the space–time approximation is second-order far from the discontinuity and first-order (in L^1) at the discontinuity vicinity. Due to the first-order accuracy ($\alpha = 1$) around the discontinuity, it is mandatory that both spatial step and time step to be four times smaller ($k = 4$) to get a four times smaller error. In other words, starting from a solution u_N obtained with a total of N degrees of freedom, the approximation error will be four times smaller only with a total number of degrees of freedom of $4^d N = 256 N$:

$$||u - u_{256N}|| \leq \frac{1}{4} ||u - u_N||. \qquad [2.2]$$

On the contrary, we could want to recover the maximal convergence order of the numerical scheme, here order two, on this discontinuous solution, that is, to use only $2^4 N = 16N$ for an error again four times smaller. To obtain this, it is necessary to concentrate the mesh *adaptively* near the discontinuity with a space–time mesh. The space–time meshing option is generally not considered in the literature because it needs, among many disadvantages, to mesh in 4D and to reformulate the numerical scheme. In contrast to space–time meshing, we restrict ourselves to time-advancing schemes, which means that the space–time mesh is constrained to be built from constant–time plans. For the same accuracy, the number of degrees of freedom of time-advancing approximations maybe larger than for space–time meshing. To reduce the time error by a factor of four in our example, the number of time levels needs to be four times larger. This factor four will multiply the spatial mesh refinement ratio.

A barrier limiting the convergence order in the presence of singularities at a maximum of $8/5$ written as

$$||u - u_{kN}|| \leq k^{-8/5d}||u - u_N||. \qquad [2.3]$$

has been identified in lemma 2.9 in Volume 1.

2.2.2. *Representativity of the spatial interpolation error*

We assume now that satisfying the usual explicit stability condition implies that the time error is of the same size as the space error. We discuss the validity of this assumption in section 2.7. With the above assumption, we are equipped with an error field at each time level, which then discretizes a space–time error field. Let us discuss which space–time norm we want to minimize. Let us compare between minimizing the error in the norm in $\mathbf{L}^q(0, T; \mathbf{L}^p(\Omega)), 1 \leq q < \infty$ or in $\mathbf{L}^\infty(0, T; \mathbf{L}^p(\Omega))$.

The *first option* $\mathbf{L}^q(0, T; \mathbf{L}^p(\Omega))$ would be coherent with the steady-state multiscale L^p theory of Chapter 5 of Volume 1, but taking it would imply that the space–time mesh can be updated only after the computation of the whole simulation time frame, in order to control the global space–time complexity or the global space–time error in $\mathbf{L}^q(0, T; \mathbf{L}^p(\Omega))$. Such a global approach is discussed in Chapter 7 of this volume.

As concerns the *second option* of $\mathbf{L}^\infty(0, T; \mathbf{L}^p(\Omega))$, we have seen in Chapter 4 of Volume 1 that $\mathbf{L}^\infty$-based adaptation does not produce second-order convergence. But since we are applying spatial adaptation, the $\mathbf{L}^\infty(0, T; \mathbf{L}^p(\Omega))$ approach will be able to spatially capture discontinuities and the error $\mathbf{L}^p$ norm of spatial approximation at each time level can be driven to small values. The $\mathbf{L}^\infty$ in time strategy consists of requirement that each mesh fulfills a specified $\mathbf{L}^p$-error=ε criterion. Adaptative spatial order is maintained at level 2 as in the first option, while, as for the first option, time accuracy constraint leads to divide the timestep by 4, as for the first option, so that the 8/5 limit can also be attained. We then feel free to select the $\mathbf{L}^\infty(0,T;\mathbf{L}^p(\Omega))$ *space–time adaptation criterion,* which we could apply as in Algorithm 2.1.

Algorithm 2.1. Transient fixed-point mesh adaptation with a new mesh every time step

---

– choose *an error threshold ε*;

– for each time level:

- compute metric $\mathcal{M}_{\mathbf{L}^p}$ for error threshold ε by applying relation [4.40] of Chapter 4 of Volume 1,

- adapt the mesh with respect to $\mathcal{M}_{\mathbf{L}^p}$,

- define the new time step length by a Courant-like criterion.

Now, Algorithm 2.1 is not satisfying. Indeed, an algorithm changing the mesh at each solver time step in order to adapt it to the solution evolution has two bottlenecks. First, it increases drastically the cost in terms of mesh generation. Second, it introduces a large amount of error due to the transfers of the solution field from one mesh to another as the new adapted mesh is not a simple deformation of the first one. These issues are motivations to *use a spatial mesh during several time steps*.

2.3. Transient fixed-point mesh adaptation scheme

The simulation time frame $[0,T]$ is split into several subintervals:

$$[0,T] = [0 = t_0, t_1] \cup ... \cup [t_i, t_{i+1}] \cup ... \cup [t_{n-1}, t_n = T].$$

The whole algorithm is presented in Algorithm 2.2. and depicted in Figure 2.1.

The external loop goes from the first subinterval to the last one. On any subinterval $[t_i, t_{i+1}]$ is applied a fixed-point iteration, combining alternatively solution computation and mesh regeneration, which corresponds to the internal loop shown in Figure 2.1.

In the internal loop, knowing the spatial mesh, the time subinterval $[t_i, t_{i+1}]$ is divided into m time-integration intervals $[t_i^k, t_i^{k+1}]$, with $k = 0, ..., m$ and $t_i^0 = t_i$, $t_i^m = t_{i+1}$. For t_i^0, the solution is given by interpolation from a previous mesh, and for any $k = 0, ..., m-1$, the flow variables are advanced from time level t_i^k to time level t_i^{k+1} by means of the numerical scheme. Then, a single metric for the whole subinterval $[t_i, t_{i+1}]$ can be defined by intersecting for all $k = 0, ..., m$ the metrics associated with solutions at time level t_i^k. In practice, not all metrics are intersected, but a few tens of them. The metric intersection procedure is explained in section 3.6 of Volume 1. Then, a new adapted mesh is generated according to this metric and to the error threshold ε. Then either the computation starts again from the same interpolation of the solution obtained at the end of previous subinterval, or this internal loop is stopped, according to a measure of the iterative convergence, namely when the deviation between two successive solutions at t_{i+1} is sufficiently small, as for the steady case.

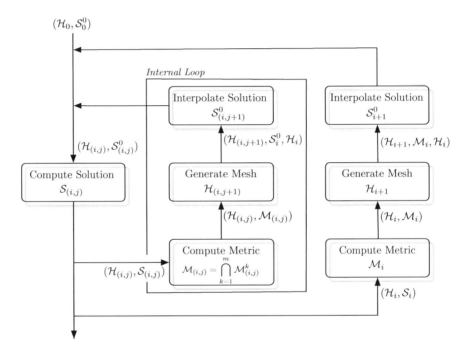

Figure 2.1. *Schematic presentation of the transient fixed-point mesh adaptation algorithm. Symbols $\mathcal{H}$, $\mathcal{S}$ and $\mathcal{M}$ stand for, respectively, mesh, flow solution and metric. The internal loop applies the fixed point process that ensures after its convergence that the spatial mesh used for advancing from t_i to t_{i+1} is adapted to any intermediate flow. The external loop organizes the transition from the end of interval $[t_i, t_{i+1}]$ to the beginning of $[t_{i+1}, t_{i+2}]$*

The next action is performed by the *external loop* that starts the next internal fixed point adaptation loop for $[t_{i+1}, t_{i+2}]$.

Algorithm 2.2. Transient $\mathbf{L}^\infty(0,T;\mathbf{L}^p(\Omega))$ fixed-point mesh adaptation algorithm

```
//- Loop over time subintervals i = 1, n_adap
```
For $i=1,n_{adap}$
    ```//- Solve adaptively on time subintervall``` $S_i = [t_{i-1}, t_i]$
    ```//- Fixed point adaptation loop```
    For $j=1,n_{nptfx}$

- $\mathcal{W}_{0,i}^j = \texttt{ConservativeSolutionTransfer}(\mathcal{H}_{i-1}^j, \mathcal{W}_{i-1}^j, \mathcal{H}_i^j)$

- $\mathcal{W}_i^j = \texttt{SolveStateForward}(\mathcal{W}_{0,i}^j, \mathcal{H}_i^j)$

- $\mathcal{M}_i^j = \texttt{ComputeFeatureOrientedMetric}(\varepsilon, \mathcal{W}_i^j, \mathcal{H}_i^j)$

- $\mathcal{H}_i^{j+1} = \texttt{GenerateAdaptedMeshe}(\mathcal{H}_i^j, \mathcal{M}_i^j)$

    End for j
End for i

For each subinterval, a maximal metric $|\mathbf{H}_{\max}|_i^j$ is computed from the different variables and different time levels of the subinterval by applying the *metric intersection* defined in section 3.6.1.

The number n_{adap} of *adaptation time intervals* $[t_i, t_{i+1}]$ is an extra discretisation parameter to be specified. Let us again consider the case of a discontinuity traveling across the spatial domain, from one side to the opposite one.

If we compute with a unique anisotropic adaptation time interval ($n_{adap} = 1$), that is, $[t_0, t_1] = [0, T]$, the final unsteady calculation is performed on one mesh, adapted to all intermediate positions of the discontinuity, which, in this particular example, results in a uniform fine mesh. In that case, the mesh adaptation cycles do not result in a better efficiency. Instead, convergence remains at first order or less.

Conversely, if the number n_{adap} of adaptation time intervals $[t_i, t_{i+1}]$ is too large, this results in (1) the costly generation of a very large number of meshes and (2) too many transfers between adapted meshes resulting in the accumulation of a large error. Therefore, for a single mesh adaptive calculation, a compromise needs to be found between the above limiting options. In the numerical example, we shall define a compromise that can be used.

2.3.1. *Size of subintervals in a mesh convergence*

A second question is: how these adaptation time intervals should be chosen in a strategy of several calculations for *mesh convergence* to continuous solution. To illustrate the impact of the time-interval size on the adapted mesh, we consider the advection by a constant velocity a discontinuity. Figure 2.2 shows the adapted mesh for the time period $[t_i, t_{i+1}]$ (on both pictures), the mesh use for the travel of a circle-shaped discontinuity. When advected, the circular discontinuity travels only in the refined regions.

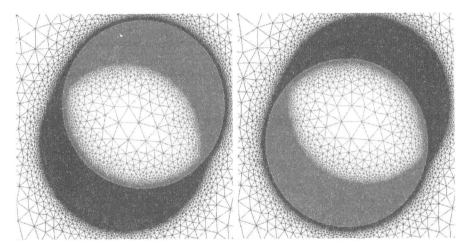

Figure 2.2. *Transient fixed-point mesh adaptation algorithm applied to the advection of a circle-shaped discontinuity between a region where physical density is equal to 1 (blue) a region where it is equal to 2 (orange). Left, the interface at time t_i. Right, the interface at time t_{i+1}. On both pictures, the adapted mesh for advection during the time interval $[t_i, t_{i+1}]$. For a color version of this figure, see www.iste.co.uk/dervieux/meshadaptation2*

Let us assume that we keep the same subinterval $[t_i, t_{i+1}]$ for the different computations of a mesh convergence. In order to reduce the error by a factor of four in 3D, we wish to divide the error by a factor of four by performing two successive computations:

– one with a coarse Mesh 1 with a mesh size of N nodes;

– one with a fine Mesh 2 with a mesh size of $8N$ nodes.

The finely discretized vicinity of the discontinuity, which is defined by the trajectory of it during the time interval, will cover the same region of the computational domain as for the coarse computation. Then, if the *isotropic* mode is chosen, the number of nodes inside this "vicinity of discontinuity" will pass from N_d for Mesh 1 to $64N_d$ for Mesh 2 since the local Δx is divided by 4. This should be $8N_d$ for second-order convergence. In this example, second-order spatial convergence is clearly unreachable. The region of high refinement swept by the discontinuity needs to be reduced by a factor of eight, which shows that the adaptation time intervals $[t_i, t_{i+1}]$ should be taken eight times smaller for the calculation with Mesh 2. In the *anisotropic* mode, the situation is much better, since running for Mesh 2 with the same adaptation time intervals as Mesh 1 would result in passing from N_d for Mesh 1 to a Mesh 2 in which, in a favorable case, only the mesh size normal to discontinuity needs to be four times smaller, while the two other directions are two times smaller, which results in a number of $16N_d$ (still larger than $8N_d$, second order). However, if in these conditions the time intervals are divided by two, then we get a number of $8N_d$:

LEMMA 2.1.– In the 3D anisotropic case, a necessary condition for second-order spatial convergence is that subintervals $[t_i, t_{i+1}]$ are *two times* smaller for a four times smaller spatial error. □

In the sequel, mesh convergence experiments will follow this rule except when mentioned.

2.3.2. *Mesh adaptation for unsteady Euler/Navier–Stokes equations with thickened interface*

2.3.2.1. *Control of interface thickening*

For simplicity, we assume in this section that the mesh is adapted at each time step. We discuss how the mesh efforts need to be balanced between the interface capture and the rest of flow resolution. Our numerical model, presented in section 1.3 of Volume 1 relies on a *thickened interface*. The sine transition with thickness η is introduced in section 1.3.3.2 of Volume 1 for spatial stability purpose. This transition does not need to be computed with a high accuracy. But the thickness η itself introduces an error with respect to the non-discretized solution, which has an interface with zero thickness. The error introduced by interface thickening is a $O(\eta)$ term. Therefore, the parameter η should tend to zero at convenient rate when the number of vertices N is increased or, equivalently, when the error parameter ε is decreased for convergence toward the continuous solution. For a convergence order α, η should satisfy:

$$\eta(N) = Cst\, N^{-\frac{\alpha}{d}} \quad \text{or} \quad \eta(\varepsilon) = Cst\, \varepsilon.$$

Here, we are interested in being close to second-order convergence. To this end, dividing the global approximation error by a factor of four requires that we multiply the mesh size by two or, equivalently, the mesh density by a factor of 2^d in smooth regions of the domain, while we need to divide the thickness η by at least a factor of four. Therefore, choosing the interface thickening η proportional to the specified error level ε provides an equilibrated error behavior in the sense that dividing ε by a factor of four gives a global error divided by four. In practice, we start with a quasi uniform mesh and we set η to a size of 2-3 space steps Δx. In the case of a fixed interface, the convergence is analyzed as follows:

LEMMA 2.2.– Let us assume that the interface does not move during the time interval. Then, the proposed approach with $\eta(\varepsilon) = Cst\ \varepsilon$ has a second-order spatial convergence rate for anisotropic adaptation while the order is at most $d/(d-1)$ for the isotropic one[1]. □

While the thickness coefficient η controls the interface thickness, the thickness of mesh refinement near interface is much larger, due to the freezing of mesh on each time subinterval. We assume that the extra thickness due to η can be neglected. Then lemma 2.1 can still be applied.

2.3.2.2. *Local criterion for bi-fluid flows*

We now discuss a heuristic choice of criterion designed to accurately capture the interface when a level set method is applied. In the context of bi-fluid flows, it is important to specify which quantities – the features – are going to govern the mesh adaptation according to the error analysis introduced in Chapter 5 of Volume 1.

First, the mesh is adapted to compute accurately the dynamic of the flow. Indeed, capturing small-scale details can be determinant for the final overall accuracy. To this

---

[1] Let us give a short proof. Indeed, we start from a pair (ε, η) for which the adaptation to the thick interface requires N_i vertices and the adaptation to the rest of the domain, the region complementary to the thick interface where the solution is smooth, necessitates N_s vertices. Then, error threshold ε and interface thickness η are divided by four. The numbers of nodes become, respectively, N'_i and N'_s. The new smooth solution area made free by passing to a thinner interface needs a number of vertices, which is negligible with respect to the rest of the domain. Thus, the new regular region needs 2^d times more vertices for an error four times smaller, according to the smooth-adaptation analysis of the previous sections ($\frac{N'_s}{N_s} = 2^d$). Let us analyze the amplification $k = N'_i/N_i$ of the number of nodes in the vicinity of the interface: thanks to the option $\eta(\varepsilon) = Cst\ \varepsilon$, the number of mesh layers close to interface is not changed, because the mesh size normal to the interface is four times smaller, but it is applied also in a width four times smaller. The mesh is also adapted in the direction tangent to the interface. The necessary number of nodes depends on whether the mesh adaptation is anisotropic or not. In the isotropic case, the volume to mesh is four times smaller, with a mesh step four times smaller. The number of nodes is multiplied by a factor: $k_{iso} = \frac{N'_{i,iso}}{N_{i,iso}} = \frac{4^d}{4} = 4^{d-1}$. In the anisotropic case, due to the bound concerning the interface curvature and measure, we can apply a two times smaller step in tangent direction(s), which, taking into account that the number of normal layers is constant, gives a factor $k_{aniso} = \frac{N'_{i,aniso}}{N_{i,aniso}} = 2^{d-1}$, less than $\frac{N'_{s,aniso}}{N_{s,aniso}}$.

end, the momentum length $\rho|\mathbf{U}|$ is used as an adaptive variable and its interpolation error is controlled in $\mathbf{L}^2$ norm. This provides $\mathcal{M}_{\mathbf{L}^2}(\rho|\mathbf{U}|)$:

$$\mathcal{M}_{\mathbf{L}^2}(\rho|\mathbf{U}|) = D_{\mathbf{L}^2} \left(\det \left|H_{\rho|\mathbf{U}|}\right|\right)^{-\frac{1}{4+d}} \left|H_{\rho|\mathbf{U}|}\right|$$

$$\text{with } D_{\mathbf{L}^2} = \frac{d}{\varepsilon} \left(\int_\Omega (\det \left|H_{\rho|\mathbf{U}|}\right|)^{\frac{2}{4+d}}\right)^{\frac{1}{2}}.$$

Second, the mesh has to be adapted to the thickened interface to control the error in capturing it. Moreover, it is important to accurately represent the interface curvature for solution accuracy. Two strategies have been considered. As the interface represents a jump of the density, or a sine variation of the density in the case of a thickened interface, a first approach is to adapt, according to Chapter 5 of Volume 1, the mesh to the density: $\mathcal{M}_{ls} = \mathcal{M}_{\mathbf{L}^2}(\rho)$. Although it performs well, this approach has the following weaknesses:

– for thickened interfaces, the density Hessian is zero in the middle of the sine variation leading to a coarser mesh in this area;

– it does not take into account the curvature of the interface;

– this adaptation is highly dependent on the specified thickness η.

We thus propose a second strategy where a metric is specifically dedicated to the interface. It is based on a metric that follows the iso-lines of the given field.

Figure 2.3. *Adaptation to the interface, that is, to the 0 value of a level set function ϕ, with the metric given by relation [2.4]. Left, the level set (iso-lines) representation of the function ϕ. Right, the mesh adapted only to the interface*

The metric $\mathcal{M}_{ls}$ is dedicated to the interface, that is, only to the 0 contour of the level set function ϕ, which is derived from the iso-lines metric $\mathcal{M}_{isolines}$. This metric for the interface is defined by four parameters:

- h_{ls} the prescribed size for the direction **n** normal to the interface;
- δ_{ls} the thickness of the adapted region on both sides of the interface;
- ε_{ls} the error threshold for the variation of the interface that controls the size prescription in the tangential direction **t**;
- h_{grad} the growth parameter of the mesh while gradually moving further from the interface.

It is given by

$$\mathcal{M}_{ls}(\mathbf{x}) = \begin{cases} \mathcal{R}(\mathbf{x}) \begin{pmatrix} h_{ls}^{-2} & 0 \\ 0 & D_{\mathbf{L}^2} \left(\det |G_\phi(\mathbf{x})| \right)^{-\frac{1}{4+d}} \gamma_{\mathbf{t}}(\mathbf{x}) \end{pmatrix} {}^t\mathcal{R}(\mathbf{x}) & \text{if } |\phi(\mathbf{x})| \leq \delta_{ls} \\ \mathcal{R}(\mathbf{x}) \begin{pmatrix} h_{\mathbf{n}}^{-2}(\mathbf{x}) & 0 \\ 0 & h_{\mathbf{t}}^{-2}(\mathbf{x}) \end{pmatrix} {}^t\mathcal{R}(\mathbf{x}) & \text{if } |\phi(\mathbf{x})| > \delta_{ls} \end{cases} \quad [2.4]$$

with the notation

$$G_\phi(P) = \frac{1}{|\mathcal{B}(P)|} \begin{pmatrix} \int_{\mathcal{B}(P)} \left|\frac{\partial \phi}{\partial x}\right|^2 & \int_{\mathcal{B}(P)} \frac{\partial \phi}{\partial x}\frac{\partial \phi}{\partial y} \\ \int_{\mathcal{B}(P)} \frac{\partial \phi}{\partial x}\frac{\partial \phi}{\partial y} & \int_{\mathcal{B}(P)} \left|\frac{\partial \phi}{\partial y}\right|^2 \end{pmatrix} = \mathcal{R} \begin{pmatrix} \gamma_{\mathbf{n}} & 0 \\ 0 & \gamma_{\mathbf{t}} \end{pmatrix} {}^t\mathcal{R} = \mathcal{R}\Gamma\,{}^t\mathcal{R}$$

and with

$$h_{\mathbf{n}}(\mathbf{x}) = (|\phi(\mathbf{x})| - \delta_{ls}) \log(h_{grad}) + h_{ls},$$
$$h_{\mathbf{t}}(\mathbf{x}) = (|\phi(\mathbf{x})| - \delta_{ls}) \log(h_{grad}) + D_{\mathbf{L}^2} \left(\det |G_\phi(\mathbf{x})| \right)^{-\frac{1}{4+d}} \gamma_{\mathbf{t}}(\mathbf{x}).$$

The reader can refer to Alauzet (2009) for details on the derivation of $h_{\mathbf{n}}(\mathbf{x})$ and $h_{\mathbf{t}}(\mathbf{x})$. Figure 2.3 displays the adaptation to the interface on the previous example of a breaking water column impacting an obstacle. The level set function ϕ is shown on the left and the adapted mesh generated from the interface metric [2.4] on the right.

To sum up, in the context of bi-fluid flows, a metric for the dynamic of the flow and a metric designed for the interface are computed. Both metrics are taken into account because of the *metric intersection* defined in section 3.6.1 of Volume 1:

$$\mathcal{M} = \mathcal{M}_{\mathbf{L}^2}(\rho|\mathbf{U}|) \cap \mathcal{M}_{ls}.$$

2.3.3. *Convergent transient fixed-point*

The *convergent* transient fixed-point also controls mesh refinement in order to converge to the exact solution. This is done through a gain factor $R > 1$ prescribed by the user. Before re-iteration of the outer loop, according to the choice between complexity controlling or error controlling, either

– the mesh complexity N_m ($\approx$ number of vertices) is increased to $N_{m+1} = R^d N_m > N_m$, that is, the mesh size is divided by a factor $R = (N_{m+1}/N_m)^{1/d} > 1$, or

– the error ε is divided by a factor R^2.

In practical cases, we choose $R = 2^{1/d}$ (doubling the complexity). We have summarized the above options in Algorithm 2.3.

2.4. 2D bi-fluid example

We present now a convergence study of the 2D version of the proposed adaptive algorithm applied to the complete bi-fluid numerical model presented in section 1.3 of Volume 1 (Niceflow platform) in the context of a bi-fluid flow simulation.

A rectangular water column is falling inside a rectangular box. Measurements have been done by Koshizuka et al. (1995). We compare the convergence of the simulation on structured uniform meshes and adaptive meshes on four outputs that are the position of the interface front at bottom of the box at different times. The first strategy uses a series of embedded structured uniform meshes of sizes between 2,500 and 40,000 vertices. The second strategy uses two outer iterations of Algorithm 2.3. with isotropic meshes and involves two adaptive simulations with error threshold 2ε and ε ($R = \sqrt{2}$). The simulation time frame is split into 40 subintervals of 0.025 s for mesh adaptation. As the number of vertices varies for the adaptive simulations, the mean number of vertices of the whole simulation, respectively, 1,800 and 2,700 vertices, is considered for the spatial convergence study. The numerical values of the four outputs of this study are summarized in Table 2.1. The performance in convergence of the calculations is sketched in Figure 2.5. Rather surprisingly, an approximatively second-order numerical convergence is observed for the three uniform mesh computations. This allows to apply a Richardson-type second-order extrapolation to the two finest calculations, giving a probably very accurate estimate of the continuous solution. On the basis of this accurate estimate, we observe that (1) adaptive computations show a higher order spatial convergence (the convergence order is observed with the thickened interfaces), at least on this time interval, and (2) the adaptive computation with an average of 2,700 vertices on one part and the uniform mesh with 40,000 on the other part produce results of same quality.

Algorithm 2.3. Convergent transient fixed-point mesh adaptation algorithm for flow with interface

```
//- Loop over mesh fineness m = 1, m_adap
```
For m=1,m_{adap}
```
  //- Loop over time subintervals i = 1, n_adap
```
  For i=1,n_{adap}
```
    //- Solve adaptively on time subinterval S_i = [t_{i-1}, t_i]
    //- Fixed point adaptation loop
```
    For j=1,n_{nptfx}

- $\mathcal{W}_{0,i}^{j} = \texttt{ConservativeSolutionTransfer}(\mathcal{H}_{i-1}^{j}, \mathcal{W}_{i-1}^{j}, \mathcal{H}_{i}^{j})$
- $\mathcal{W}_{i}^{j} = \texttt{SolveStateForward}(\mathcal{W}_{0,i}^{j}, \mathcal{H}_{i}^{j})$
- $\mathcal{M}_{i,u}^{j} = \texttt{ComputeFlowMetric}(\varepsilon_m, \mathcal{W}_{i}^{j}, \mathcal{H}_{i}^{j})$
- $\mathcal{M}_{i,\gamma}^{j} = \texttt{ComputeInterfaceMetric}(\varepsilon_m, \mathcal{W}_{i}^{j}, \mathcal{H}_{i}^{j})$
- $\mathcal{M}_{i}^{j} = \texttt{ComputeMetricIntersection}(\varepsilon_m, \mathcal{M}_{i,u}^{j}, \mathcal{M}_{i,\gamma}^{j})$
- $\mathcal{H}_{i}^{j+1} = \texttt{GenerateAdaptedMeshe}(\mathcal{H}_{i}^{j}, \mathcal{M}_{i}^{j})$

  End for j

End for i
```
Refinement factor R > 1:
Decrease error ε_m → ε_{m+1} = R^{-2} ε_m  or complexity  N_{m+1} = R^d N_m > N_m
Decrease interface thickness η_m → η_{m+1} = R^{-1} η_m
Decrease subinterval sizes s_m → s_{m+1} = R^{-1} s_m
Re-use the first metric.
```
End for m

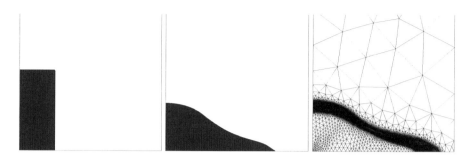

Figure 2.4. *2D falling water column. Interfaces at $t = 0$ and $t = 2$ and mesh at $t = 2$. For a color version of this figure, see www.iste.co.uk/dervieux/meshadaptation2*

Mesh Dimensionless time	0.5	1	1.5	delta/order
S1 Structured (2,500 vertices)	0.732	0.939	1.374	0.093
S2 Structured (10,000 vertices)	0.736	0.982	1.442	0.025
S3 Structured (40,000 vertices)	0.740	1.013	1.461	0.006/2.0
S2-S3 Second-order extrapolation	0.7413	1.022	1.467	0.
Adaptive (mean size $\approx$ 1,800 vertices)	0.735	0.994	1.45	0.017
Adaptive (mean size $\approx$ 2,700 vertices)	0.741	1.019	1.46	0.007/2.1

Table 2.1. *2D falling water column. The x-location of the bottom of the column at different dimensionless time ($t\sqrt{2g/L_{ref}}$). Spatial convergence is evaluated with three embedded structured uniform meshes (lines 1-3), an extrapolation from these (line 4) and two adaptive mesh simulations (lines 5-6)*

2.5. Example: impact of a 3D water column on a obstacle

This three-dimensional example is presented in detail in Guégan et al. (2010). It aims at validating the proposed method in a long-time simulation involving a 3D complex interface. We use isotropic meshes[2]. The problem consists of a water column falling in a parallelepipedic box containing a cubic obstacle. This experiment has been performed by the Maritime Research Institute Netherlands (MARIN)[3]. Water height and pressure measurements are available on a series of points as functions of time. Their positions in the computational domain are shown in Figure 2.6. All test case and experiment data are available on the Smoothed Particle Hydrodynamics European Research Interest Community[4] (SPHERIC). The

---

2 Anisotropic unsteady calculations are described in Chapter 7 of this volume.
3 Available at: http://www.marin.nl/web/show.
4 Available at: http://wiki.manchester.ac.uk/spheric/index.php/SPHERIC_Home_Page.

experiment involves a violent transient flow with a very complex interface when the water impacts the obstacle and the opposite wall (at a physical time close to 2 s). Then, the flow returns to a smooth sloshing mode. Several calculations of this case have been presented in the literature (see, for instance, Elias and Coutinho (2007) and Kleefsman et al. (2005)). They illustrate that long-term accuracy is a difficult challenge. The simulation described here[5] has been run during a physical time of 6 s, which corresponds to a forward wave motion, a backward one, and then a second forward motion. The mesh adaptation is chosen to be isotropic and follows Algorithm 2.2. The simulation time interval has been split into 120 subintervals of 0.05 s. The interface evolution obtained in this simulation is depicted in Figures 2.7 and 2.8 at different physical times. It is compared to pictures of the MARIN experiment[6] (on the right).

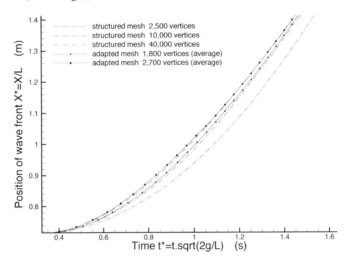

Figure 2.5. *2D falling water column. The x-location of the bottom of the column is presented as a function of time during the early acceleration of the process. Three embedded structured uniform mesh computations and two adaptive mesh computations are compared. See Table 2.1 and text for analysis*

---

5 Code NiceFlow in MPI mode.

6 It is important to note that in MARIN pictures only a part of the domain is represented. The part of the domain where the water was initially held back by the hatch is missing. It represents one-third of the domain total length. This missing part is shown by the icon top right of the picture.

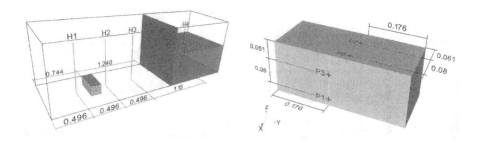

Figure 2.6. *3D falling water column on a obstacle. Left, the simulation geometry with the initial conditions and the position of the water height sensors. Right, the position of the pressure sensors on the obstacle. Pictures courtesy of R.N. Elias and A.L.G.A. Coutinho extracted from Elias and Coutinho (2007)*

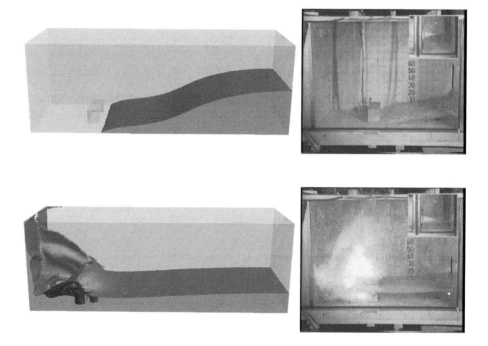

Figure 2.7. *3D falling water column on a obstacle. Comparison between the interface obtained in the simulation (left) and the pictures from the MARIN experiment (right). From top to bottom, snapshots for every 0.4 s, for times $t = 0.4$ s and $t = 1.6$ s. For a color version of this figure, see www.iste.co.uk/dervieux/meshadaptation2*

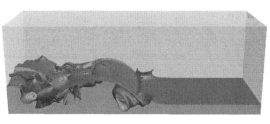 

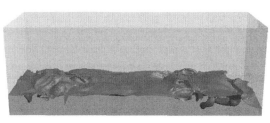 

Figure 2.8. *3D falling water column on a obstacle. Comparison between the interface obtained in the simulation (left) and the pictures from the MARIN experiment (right). From top to bottom, snapshots for $t = 2.4$ s and $t = 5.6$ s. For a color version of this figure, see www.iste.co.uk/dervieux/meshadaptation2*

The interface geometry at physical times 1.6 s and 2.4 s demonstrates the complexity of the simulation, notably by the presence of several tube- and veil-shaped structures for the interface. With a capillarity model, these structures would transform into drops, closer to the physics. The bottom picture of Figure 2.8 shows the return to equilibrium of the flow at time 5.6 s.

An example of the meshes used to advance the solution is presented in Figure 2.9 (left). Figure 2.9 (right) plots the evolution of the number of vertices with respect to the physical time. A detailed comparison with experimental measures is presented in Guégan et al. (2010).

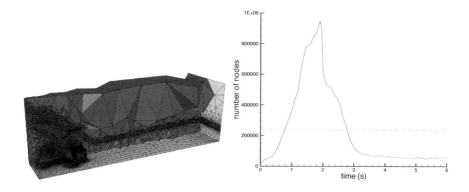

Figure 2.9. *3D falling water column on a obstacle. Mesh adaptation based on the interface and moments. Left: An example of mesh used during a time subinterval, around $t = 1.2s$. The mesh involves $\approx$ 500,000 vertices. Right: Variation of the number of mesh vertices as a function of time. The dashed line represent the average number of vertices for the whole simulation $\approx$ 240,000 vertices. For a color version of this figure, see www.iste.co.uk/dervieux/meshadaptation2*

2.6. Conclusion

This chapter presents a first transient mesh adaptation algorithm. A second algorithm is presented in Chapter 7 of this volume. Since mesh adaptation is built from a fixed point, mesh adaptation is time implicit, that is, stable and by construction adapted to the new time levels. The adaptation is designed from a control of interpolation error in momentum field together with interface representation error. For interface error control, two criteria have been introduced and discussed. Taking into account the level set curvature allows for an improved representation of corners and small details. A particular attention has been paid to the design of an adaptation algorithm with a higher order convergence rate. Time dimension is taken into account by the choice of a constant-error criterion. This means that when the flow becomes more complex with time, so does the mesh, the final accuracy being the priority.

2.7. Appendix: remarks about the adaptation of the time step

We have not addressed the complex question of the adaptation of the timestep. In many of our applications, the model is advection dominated and advanced with explicit time stepping. In that case, a simplified assumption is as follows:

The time discretization error is approximatively bounded by a constant times the space approximation error.

Therefore, minimizing the spatial error implies that the time error is somewhat minimized accordingly.

Let us give a rather heuristic proof that the above assumption is essentially verified. We analyze the error in time obtained with a first-order accurate scheme with an explicit discretization in time on the transport equation (a first-order hyperbolic problem) in one dimension:

$$u_t + c\, u_x = 0, \qquad [2.5]$$

with $c > 0$. Let x_j for $j = 1, ..., N$ be a uniform spatial discretization of the one-dimensional domain and let t_n for $n = 0, ..., T$ be a uniform time discretization. For each vertex x_j at time t_n, we have $u_j^n \approx u(x_j, t_n)$. The first-order upwind method is written

$$\frac{u_j^{n+1} - u_j^n}{\Delta t} + c\, \frac{u_j^n - u_{j-1}^n}{\Delta x} = 0. \qquad [2.6]$$

Using Taylor expansion, the local truncation error τ_j^n is given by

$$\tau_j^n = c\, \frac{\Delta x}{2}\, u_{xx}(x_j, t_n) - \frac{\Delta t}{2}\, u_{tt}(x_j, t_n) + \mathcal{O}(\Delta x^2, \Delta t^2).$$

An estimation of the spatial and the time error are provided by the first and the second term, respectively (indices are omitted for clarity): $\varepsilon(\Delta x) = c\frac{\Delta x}{2}\, u_{xx}$ and $\varepsilon(\Delta t) = -\frac{\Delta t}{2}\, u_{tt}$. Applying twice the continuous equation, we get $u_{tt} = c^2\, u_{xx}$, therefore $\varepsilon(\Delta t) = -c^2 \frac{\Delta t}{2}\, u_{xx}$. With CFL condition $\nu = c\, \frac{\Delta t}{\Delta x} \leq 1$, this implies that the time error is bounded by the spatial error

$$|\varepsilon(\Delta t)| \leq |c\, \frac{\Delta x}{2}\, u_{xx}| = |\varepsilon(\Delta x)|.$$

This analysis extends to higher order schemes, in particular to the second-order schemes used in this book. We refer to Alauzet et al. (2007) for a more complete analysis.

2.8. Notes

The transient fixed-point mesh adaptation method was introduced in Alauzet et al. (2007). More examples of computations can be found in Guégan et al. (2010).

Dynamic adaptation with unstructured meshes dates back to previous works (Löhner 1989). Some bibliography can be found in Cao et al. (2003). Let us also refer to the monograph *Adaptive Moving Mesh Methods* by Huang and Russel (2011).

3

Multi-rate Time Advancing

We have seen that for steady problems, flows with singularities can be computed with second-order convergence by mesh adaptative strategies, that is, the best possible convergence when scheme's truncation is of second order. We have pointed out that mesh-adaptative strategies combined with usual time advancing seem to have a *convergence barrier* limiting the convergence order to a number notably smaller than two. One way to recover the convergence order partly or totally is through multi-rate time stepping. This chapter focuses on showing this property and describing a particular multi-rate method applying easily to unstructured finite-volume approximation. We discuss the details of its implementation, in particular with massive parallelism. Two examples show that, even without mesh adaptation, the method increases efficiency.

3.1. Introduction

A frequent configuration in CFD calculations combines an explicit time-advancing scheme for accuracy and a computational grid with a very small portion of much smaller elements than in the remaining mesh.

A first example is the hybrid RANS/LES simulation of high Reynolds number flows around bluff bodies. In that case, very thin boundary layers must be addressed with extremely small cells. When applying explicit time advancing, the computation is penalized by the very small time-step to be applied (CFL number close to unity). But the boundary layer is not the only interesting region of the computational domain. An important part of the meshing effort is also devoted to large regions of medium cell size in which the motion of vortices needs to be accurately captured. In these vortical regions, the best time-step for efficiency and accuracy is of the order of the ratio of local mesh size by vortex velocity, corresponding to CFL close to unity. An option is to apply globally an implicit scheme with large CFL on boundary layers, and a local CFL close to 1 on vortical regions. However, this may have several disadvantages. First a large CFL may not allow a sufficient unsteady accuracy in the boundary layer region, due to unsteady separation, for example. Second, vortices motion need high advection accuracy. Rather simple implicit schemes using backward differencing show much more dissipation than explicit schemes, and high-order implicit schemes are complex and CPU consuming.

The second example concerns an important complexity issue in unsteady mesh adaptation. Indeed, time-leveled mesh adaptive calculations are penalized by the very small time-step imposed to the whole domain by accuracy requirements arising solely on regions involving small space–time scales. This small time step is an important computational penalty for:

– mesh adaptive methods of AMR type (Berger and Colella 1989);

– mesh adaptation by mesh motion;

– the transient fixed-point mesh-adaptive methods introduced in Chapters 2 and 7 of this volume. In the transient fixed point, the loss of efficiency is even more crucial when the anisotropic mesh is locally strongly stretched. Due to this loss, the numerical convergence order for flows involving discontinuities is limited to $8/5$ instead of second-order convergence (see Chapter 2 of Volume 1). Let us consider the travel of an isolated discontinuity. The discontinuity needs to be accurately followed by the mesh, preferably in a mesh-adaptive mode. Except if the adaptation works in a purely Lagrangian mode, an implicit scheme will smear the discontinuity of the solution and an explicit scheme will apply a costly very small time step on the whole computational domain.

In order to overcome these problems, the multi-rate time stepping approach represents an interesting alternative. A part of the computational domain is advanced in time with the small time-step imposed by accuracy and stability constraints. Another part is advanced with a larger time-step giving a good compromise between accuracy and efficiency. Multi-rate time advancing has been studied for PDEs and hyperbolic conservation laws (Constantinescu and Sandu 2007; Sandu and Constantinescu 2009; Mugg 2012; Kirby 2002; Seny et al. 2014; Löhner et al. 1984), and for a few applications in CFD (Löhner et al. 1984; Seny et al. 2014). Other references are proposed in section 3.6.

In this chapter, we describe a multi-rate scheme, which is based on control volume agglomeration. This scheme is well suited to a large class of finite volume approximations, and very simple to develop in an existing software relying on explicit time advancing. The agglomeration produces macro-cells by grouping together several neighboring cells of the initial mesh. The method relies on a prediction step where large time steps are used with an evaluation of the fluxes performed on the macro-cells for the region of smallest cells, and on a correction step advancing solely the region of small cells, this time with a small time step. We demonstrate the method for the mixed finite volume/finite element approximation introduced in Chapter 1 of Volume 1. Target applications are three-dimensional unsteady flows modeled by the compressible Navier–Stokes equations equipped with turbulence models and discretized on unstructured possibly deformable meshes. The numerical illustration involves the two above examples.

The algorithm is described in section 3.2. Section 3.3 analyzes the impact of passing to multi-rate with this method. Section 3.4 gives several examples of applications.

3.2. Multi-rate time advancing by volume agglomeration

3.2.1. *Finite volume Navier–Stokes*

The proposed multi-rate time-advancing scheme based on volume agglomeration will be denoted by MR and is developed for the solution of the three-dimensional compressible Navier–Stokes equations (Chapter 1, section 1.2 of Volume 1). The discrete Navier–Stokes system is assembled by a flux summation Ψ_i involving the convective and diffusive fluxes evaluated at all the interfaces separating cell i and its neighbors. More precisely, the finite volume spatial discretization combined with an explicit forward-Euler time-advancing is written for the Navier–Stokes equations possibly equipped with a $k - \varepsilon$ model as

$$vol_i\, w_i^{n+1} = vol_i\, w_i^n + \Delta t\, \Psi_i, \quad \forall\, i = 1, ..., ncell,$$

where vol_i is the volume of cell i, Δt is the time step and $w_i^n = (\rho_i^n, (\rho u)_i^n, (\rho v)_i^n, (\rho w)_i^n, E_i^n, (\rho k)_i^n, (\rho \varepsilon)_i^n)$ are as usually the density, moments, total energy, turbulent energy and turbulent dissipation at cell i and time level t^n, and $ncell$ is the total number of cells in the mesh. In the examples given below, the accuracy of the initial scheme can be defined as a third-order spatial accuracy on smooth meshes, through the use of a MUSCL-type upwind-biased finite volume, combined with the Shu and Osher (1988) third-order time accurate RK3-SSP multi-stage scheme given in Chapter 1 of Volume 1. Given an *explicit time advancing*, we assume that we can define a maximal stable time step (*local time step*) $\Delta t_i, i = 1, ..., ncell$ on each node. For the Navier–Stokes model, the stable local time step is defined by the combination of a viscous stability limit and an advective one according to the following formula:

$$\Delta t_i \leq \frac{CFL \times \Delta l_i^2}{\Delta l_i(\|\mathbf{u_i}\| + c_i) + 2\frac{\gamma}{\rho_i}\left(\frac{\mu_i}{Pr} + \frac{\mu_{t_i}}{Pr_t}\right)},$$

where Δl_i is a local characteristic mesh size, $\mathbf{u}_i$ is the local velocity, c_i is the sound celerity, γ is the ratio of specific heats, ρ_i is the density, $\frac{\mu_i}{Pr} + \frac{\mu_{t_i}}{Pr_t}$ is the sum of local viscosity to Prandtl ratio, laminar and turbulent, and *CFL* is a parameter depending on the time-advancing scheme of the order of unity. Using the local time step Δt_i leads to a stable but not consistent time advancing. A stable and consistent time advancing should use a *global/uniform time step* defined by

$$\Delta t = \min_{1, ncell} \Delta t_i.$$

For many advective explicit time advancing, in regions where Δt is of the order of Δt_i, accuracy is quasi-optimal, and in other regions, the accuracy is suboptimal, due to the relatively large spatial mesh size.

3.2.2. *Inner and outer zones*

We first define the inner zone and the outer zone, the coarse grid and the construction of the fluxes on the coarse grid, ingredients on which our MR time-advancing scheme is based. For this splitting into two zones, the user is supposed to choose a (integer) *time step factor* $K > 1$. We define the *outer zone* as the set of cells i for which the explicit scheme is stable for a time step $K\Delta t$:

$$\Delta t_i \geq K\Delta t,$$

and the *inner zone* is the set of cells i for which

$$\Delta t_i < K\Delta t.$$

We shall build over the whole domain a coarse grid which should allow that:

– advancement in time is performed with time step $K\Delta t$;

– advancement in time preserves accuracy in the outer zone;

– advancement in time is consistent in the inner zone.

A *coarse grid* is defined on the inner zone by applying cell agglomeration in such a way that on each macro-cell, the maximal local stable time step is at least $K\Delta t$. Agglomeration consists of considering each cell and aggregating to it neighboring cells, which are not yet aggregated to another one (Figure 3.1 and Algorithm 3.1.).

Agglomeration into macro-cells is re-iterated until all macro-cells with maximal time step smaller than $K\Delta t$ have disappeared. We advance in time the chosen explicit scheme on the coarse grid with $K\Delta t$ as time step. The nodal fluxes Ψ_i are assembled on the fine cells (as usual). Fluxes are then summed on the macro-cells I (inner zone):

$$\Psi^I = \sum_{k \in I} \Psi_k. \qquad [3.1]$$

Algorithm 3.1. A basic agglomeration algorithm

Consider successively every cell.
1) if the cell C is already included in a macro-cell, then consider next cell; otherwise, create a new macro-cell and put into this macro-cell each cell:
 - neighboring C;
 - not already included in a macro-cell;
2) if the new macro-cell contains only cell C, then destroy that macro-cell and put cell C in an existing macro-cell containing a neighbor of C;
3) next cell.

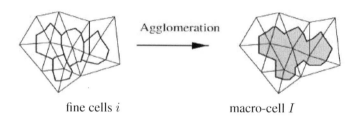

Figure 3.1. *Sketch (in 2D) of the agglomeration of four cells into a macro-cell. Cells are dual cells of triangles, bounded by sections of triangle medians*

3.2.3. *MR time advancing*

The MR algorithm is based on a *prediction step* and a *correction step* as defined hereafter:

Step 1 (prediction step):

The solution is advanced in time with time step $K\Delta t$; on the fine cells in the outer zone and on the macro-cells in the inner zone (which means using the macro-cell volumes and the coarse fluxes as defined in expression [3.1]):

For $\alpha = 1, nstep$

outer zone: $vol_i w_i^{(\alpha)} = a_\alpha vol_i w_i^{(0)} + b_\alpha \left(vol_i w_i^{(\alpha-1)} + K\Delta t\, \Psi_i^{(\alpha-1)}\right)$ [3.2]

inner zone: $vol^I w^{I,(\alpha)} = a_\alpha vol^I w^{I,(0)} + b_\alpha \left(vol^I w^{I,(\alpha-1)} + K\Delta t\, \Psi^{I,(\alpha-1)}\right)$ [3.3]

$w_i^{(\alpha)} = w^{I,(\alpha)}$ for $i \in I$ [3.4]

EndFor α,

where $w^{I,(\alpha)}$ denotes the fluid/turbulent variables at macro-cell I and stage α, (a_α, b_α) are the multi-stage parameters, $a_1 = 0, a_2 = 3/4, a_3 = 1/3$ and $b_1 = 1, b_2 = 1/4, b_3 = 2/3$ for the three-stage time advancing and vol^I is the volume of macro-cell I. From a practical point of view, we do not introduce in our software a new variable w^I to that already existing w_i. The previous sequence [3.3] is actually

inner zone: $vol^I w_i^{(\alpha)} = a_\alpha vol^I w_i^{(0)} + b_\alpha \left(vol^I w_i^{(\alpha-1)} + K\Delta t\, \Psi^{I,(\alpha-1)}\right)$, [3.5]

where I is the macro-cell containing cell i. [3.6]

On the other hand, the coarse fluxes Ψ^I are not stored in a specific variable, which means that no extra storage is necessary than the one required to store the fluxes Ψ_i[1]. Indeed, after the computation of the fluxes $\Psi_i^{(\alpha-1)}$ (using the values of $w_i^{(\alpha-1)}$) at each stage α of the multi-stage time-advancing scheme, the coarse flux $\Psi^{I,(\alpha-1)}$ is evaluated according to expression [3.1] for each macro-cell I and then stored in the memory space allocated to $\Psi_i^{(\alpha-1)}$ for each cell i in the inner zone belonging to the macro-cell I.

Step 2 (correction step):

– The unknowns in the outer zone are frozen at level $t^n + K\Delta t$.

– The unknowns in the outer zone close to the inner zone, which are necessary for advancing in time the inner zone (which means those which are useful for the computation of the fluxes Ψ_i in the inner zone), are interpolated in time.

---

[1] A more efficient implementation would not compute the fluxes between two internal cells of a macro-cell.

– Using these interpolated values for the computation of the fluxes Ψ_i in the inner zone (at each stage of the time-advancing scheme), the solution in the inner zone is advanced K times in time with the chosen explicit scheme and time step Δt.

This time advancing is written as:

For $kt = 1, K$
   For $\alpha = 1, nstep$

$$\text{inner zone: } vol_i w_i^{(\alpha)} = a_\alpha vol_i w_i^{(0)} + b_\alpha \left(vol_i w_i^{(\alpha-1)} + \Delta t\ \Psi_i^{(\alpha-1)} \right) \quad [3.7]$$

(outer zone: nothing is done)   [3.8]

   EndFor α.
EndFor kt.

The arithmetic complexity is proportional to the number of points in the inner zone. The global multi-rate algorithm is presented in Algorithm 3.2.

Algorithm 3.2. Multi-rate time-advancing algorithm

```
For time level l = 0 to l_max
//- Prediction step
   For α = 1, n_step
```

- Outer zone: $vol_i w_i^{(\alpha)} = a_\alpha vol_i w_i^{(0)} + b_\alpha \left(vol_i w_i^{(\alpha-1)} + K\Delta t\ \Psi_i^{(\alpha-1)} \right)$
- Inner zone: $vol^I w_i^{(\alpha)} = a_\alpha vol^I w_i^{(0)} + b_\alpha \left(vol^I w_i^{(\alpha-1)} + K\Delta t\ \Psi^{I,(\alpha-1)} \right)$

End for α
```
//- Correction step
   For kt = 1, K and α =, n_step
```

- Inner zone: $vol_i w_i^{(\alpha)} = a_\alpha vol_i w_i^{(0)} + b_\alpha \left(vol_i w_i^{(\alpha-1)} + \Delta t\ \Psi_i^{(\alpha-1)} \right)$

EndFor α, EndFor kt
```
End for l
```

3.3. Elements of analysis

3.3.1. *Stability*

The central question concerning the coarse grid is the stability resulting from its use in the computation. Considering [3.1], we expect that the viscous stability limit will improve by a factor four (1D) for a twice larger cell. The viscous stability limit can therefore be considered as more easily addressed by our coarsening. For the advective stability limit, we can be a little more precise. The coarse mesh is an unstructured partition of the domain in which cells are polyhedra. Analyses of time-advancing schemes on unstructured meshes are available in L^2 norm for unstructured meshes (see Angrand and Dervieux 1984 and Giles 1997b, 1987). Here, we solely propose a L^∞ analysis of the advection scheme. The gain in L^∞ stability can be analyzed for a first-order upwind advection scheme. We get the following (obvious) lemma:

LEMMA 3.1.– The upwind advection scheme is positive on the mesh made of macro-cells as soon as for all macro-cell I:

$$\Delta t \, ||V_I|| < \left[\sum_{J \in \mathcal{N}(I)} \int_{\partial cell(I) \cap \partial cell(J)} d\Sigma \right]^{-1} \int_{cell(I)} d\mathbf{x},$$

where $\mathcal{N}(I)$ holds for the neighboring macro-cells of I. □

The application of an adequate neighboring-cell agglomeration, like in Lallemand et al. (1992) producing large macro-cells of good aspect ratio, will produce a K-times larger stability limit.

3.3.2. *Accuracy*

In contrast to more sophisticated MR algorithms, the proposed method has not a rigorous control of the accuracy. Let us however remark that the generic situation involves variable-size meshes, which limits the unsteady accuracy on small-scale propagation, already before applying the MR method. However, the two following remarks tend to show that the scheme accuracy is conserved:

– the predictor step involves simply a sum of the fluxes and the formal accuracy order is kept, with a coarser mesh size;

– still during the predictor step, if we assume that the mesh is reasonably smooth, then the CFL applied in the inner part near the matching zone will be close to the explicit CFL (applied on the outer part near the matching zone) and therefore accuracy is high.

Under these conditions, the effect of the corrector step will improve the result. In practice, most of our experiments will involve a comparison between the explicit time advancing and its MR extension.

3.3.3. *Efficiency*

The proposed two-level MR depends on only one parameter, the ratio K between the large and small time step. Considering a mesh with N vertices, a short loop on the mesh will produce the function $K \mapsto N^{small}(K) \leq N$, which gives the number of cells in the inner region for K. If $CPU_{ExpNode}(\Delta t)$ denotes the CPU per node and per time step Δt of the underlying explicit scheme, a model for the MR cpu per Δt would be

$$CPU_{MR(K)}(\Delta t) = \left(\frac{N}{K} + N^{small}(K) \right) \times CPU_{ExpNode}(\Delta t),$$

to be compared with the explicit case

$$CPU_{Expli}(\Delta t) = N \times CPU_{ExpNode}(\Delta t).$$

We shall call the *expected gain* the following ratio:

$$\text{Gain} = \frac{CPU_{Expli}(\Delta t)}{CPU_{MR(K)}(\Delta t)} = \frac{1}{\frac{1}{K} + \frac{N^{small}(K)}{N}}.$$

The above formula emphasizes the crucial influence on efficiency of a very small proportion of inner cells.

REMARK 3.1.– In most other multi-rate methods, the phase with a larger time-step does not concern the inner region and then their gain would be modeled by

$$\text{Gain} = \frac{1}{\frac{1}{KN}(N - N^{small}(K)) + \frac{N^{small}(K)}{N}}.$$

Both gains are bounded by $N/N^{small}(K)$ and show that this ratio has to be sufficiently large.

REMARK 3.2.– Once we have evaluated $K \mapsto N^{small}(K)$ for a given mesh, it is possible to predict a theoretical optimum K_{opt} for minimizing the CPU time in *scalar* execution. However, we shall see that the pertinence of the above theory is lost by *parallel* implementation conditions.

3.3.4. *Toward many rates*

Multi-rate strategies are supposed to extend to more than two different time step lengths while keeping a reasonable algebraic complexity. Let us examine the case of *three* lengths, namely Δt, $K\Delta t$, $K^2 \Delta t$. It is then necessary to generate two *nested* levels of agglomeration in such a way that $Grid1$ is stable for $CFL = 1$, $Grid2$ is stable for $CFL = K$ and $Grid3$ is stable for $CFL = K^2$. While a *two-rate* calculation involves a prediction-correction based on $Grid1$ and $Grid2$,

– prediction on $Grid2$;

– correction on inner part of $Grid1$,

in a *three-rate* calculation, the correction step is replaced by two corrections:

– prediction on $Grid3$;

– correction on medium part of $Grid2$;

– correction on inner part of $Grid1$,

but this replacement is just the substitution (on a part of the mesh) of a single-rate advancing by a two-rate one and therefore can carry a higher efficiency (the smallest time step is restricted to a smaller inner zone). In contrast to other MR methods, we have a (second) duplication of flux assembly on the inner zone. However, this increment remains limited, since this computation is done $1 + K^{-1} + K^{-2}$ times (111% for $K = 10$).

3.3.5. *Impact of our MR complexity on mesh adaption*

We now check that the proposed MR indeed improves mesh adaption convergence order. Let us consider the space–time mesh used by a time-advancing method. A usual time advancing uses the Cartesian product $\{t_0, t_1, ...t_N\} \times$ spatial mesh as space–time mesh. The space–time mesh is a measure of the computational cost since the discrete derivatives are evaluated on each node (t_k, x_k) of the $N_{time} \times N_{space}$ nodes of the space–time mesh. In Chapter 2 of Volume 1 and Chapters 2 and 7 of this volume, an analysis is proposed, which determines the maximal convergence order (in terms of number of space-time nodes) that can be attained on a given family of mesh. This analysis is useful for evaluating mesh-adaptive methods. For example, in Chapter 2 of Volume 1 and Chapter 7 of this volume, a 3D mesh-adaptive method for computing a traveling discontinuity has a convergence order α not better than $\alpha_{max} = 8/5$, according to

$$error = O(N_{st}^{-\alpha/d}),$$

$d = 4$ being the space–time dimension.

LEMMA 3.2.– Replacing the usual time-advancing by the MR algorithm will indeed improve the maximal convergence order of a mesh adaption method as defined in Chapters 2 and 7 of this volume.

PROOF.– Similarly to previous analyses, we restrict our argument to the case of the capture of a field involving a moving discontinuity. We have seen that anisotropic adaptation is able to generate a sequence of meshes for which eight times more vertices allows a four-times-smaller spatial error in the capture of the discontinuous function (spatial second-order). We have seen in Chapter 2 of Volume 1 that this property extends to transient fixed point mesh adaptation, giving again spatial second-order. We recall that, unfortunately, this needs to be combined with a four times smaller time step, so that in space-time, the number of degrees of freedom is multiplied by $8 \times 4 = 32$, which corresponds to a space-time convergence order limited to $8/5$.

With an MR time advancing, the four times smaller time step can be restricted to the smallest spatial cells following the discontinuity. The number of points concerned is concentered along a 2D surface, so that their number can be evaluated by $N_{space}^{2/3}$ where N_{space} is the total number of spatial vertices. On finer anisotropic mesh, $N_{space}^{2/3}$ is multiplied by 4 (2D division). Then when we divide space and time, and the amplification of CPU time can be evaluated as:

$$N_{space}/\Delta t \to K N_{space}/\Delta t = 8 \times N_{space}/(\Delta t/2) + 4 \times N_{space}^{2/3}/(\Delta t/4)$$

that is with a factor:

$$K = 16 + 16 \times N_{space}^{-1/3}$$

for the proposed MR algorithm (and $16 \times (1 - N_{space}^{-1/3}) + 16 \times N_{space}^{-1/3}$ for a usual MR algorithm). Both formulas, for N_{space} large, give the second order space-time convergence. □

3.3.6. *Parallelism*

3.3.6.1. *Implementation of MR*

The proposed method needs be adapted to massive parallelism. We consider its adaptation to parallel MPI computation relying on mesh partitioning. In a preprocessing phase, the cell agglomeration is applied at run time inside each partition, which saves communications, and over the whole partition. The motivation is to do it once for the whole computation, while fluctuations of the inner zone at each time level will be taken into account by changing the list of active macro-cells to be agglomerated in the inner zone. Since our purpose is to remain with a rather simplified modification of the initial software, we did not modify the communication library (where we could to restrict the communications to the inner zone in the correction step).

To summarize, the MR algorithm involves at each time step:

– an updating of the inner zone (with a volume agglomeration done once for the computation);

– a prediction step that is similar to an explicit step (with a larger time step length), but with also a local sum of the fluxes in each macro-cell;

– a correction step that is similar to explicit arithmetics restricted to the inner region, and for simplicity of coding, communications that are left identical to the explicit advancing (communications applied to both inner and outer zones).

An intrinsic extra cost of our algorithm with respect to previous multi-rate algorithms is the computations on the inner zone during the predictor step. The correction step complexity is close to an explicit advancing one on the inner zone except the two phases of time interpolation and communications. Time interpolation can be implemented with a better efficiency by applying it only on a layer around the inner zone. However, the cost of the time interpolation is a very small part of the total cost. Global communications are less costly than 10% of the explicit time step cost. If the inner zone is 30% of the domain, developing communication restricted to inner zone will reduce the communication from 10% to 3% of the explicit time step on the whole domain, which shows that the correction step would be decreased from 40% to 33% of an explicit time stepping CPU.

3.3.6.2. *Load balancing*

The usual METIS software can be applied on the basis of a balanced partitioning of the mesh. However, as remarked in previous works (see, for example, Seny et al. 2014), if the mesh partition does not take into account the inner zone, then the work effort will not be balanced during the correction step. The bad work balance for the correction step can be of low impact if this step concerns a sufficiently small part of the mesh, resulting in a small part of the global work. However, a more reasonable assumption is that the correction phase represents a non-negligible part of the effort. In this section, we discuss the question of a partitioning taking into account the correction phase. We observe that in the proposed method the inner zone depends on the flow through the CFL condition. This means that *dynamic load balancing* may be useful in some case. It would be anyway compulsory if a strong mesh adaption like the algorithm described in Chapter 2 of this volume is combined with the multi-rate time advancing. However, in the class of flows which we consider, the change in inner zone can be neglected and we consider only static balancing. An option resulting from the work of Karypis and Kumar (2006) and available in METIS is the *multi-constraint partitioning* (MCP), which minimizes the communication cost with multiple constrains. The two constraints for MR are as follows:

– partition is balanced for the whole computational domain, which would be optimal for the prediction step;

– partition is balanced for the inner part of the computational domain, which would be optimal for the correction step.

More precisely, this partitioning algorithm produces a compromise among the following:

– making the number of nodes in each subdomain of the global mesh uniform;

– making the number of nodes in each subdomain of the inner part of the mesh uniform;

– minimizing the communications between the subdomains.

In some particular cases, the user can specify an evident partition that perfectly balances the number of nodes in each subdomain of the global mesh and in each subdomain of the inner part of the mesh. In our experiments, we explicitly specify when it is the case and how it is performed.

3.4. Applications

The MR algorithm is implemented into code Aironum. The spatial approximation is the superconvergent implementation of section 1.1.4 of Volume 1 of Chapter 1 and the circumcenter cells are used (Appendix of Chapter 1 of Volume 1). The superconvergent implementation of section 1.1.4 of Chapter 1 of Volume 1 is used. The explicit time advancing is a three-stage Shu method. The mean CPU for an explicit time step per mesh node varies between 10^{-7} and 4×10^{-7} s according to the partition quality and the number of nodes per subdomain.

3.4.1. *Circular cylinder at very high Reynolds number*

The discussion concerning the parallel processing of the multi-rate scheme will rely on a non-adaptative application. It is the simulation of the flow around a circular cylinder at Reynolds number 8.4×10^6. A hybrid RANS/VMS-LES model is used to compute this flow. The complete model is described in Itam et al. (2018). Mean flow variables and turbulent variables are advanced by the time integration scheme, and therefore also the MR method. The computational domain is made up of small cells around the body in order to allow a proper representation of the very thin boundary layer that occurs at such a high Reynolds number. Figure 3.2 depicts the Q-criterion isosurfaces and shows the very small and complex structures that need to be captured by the numerical and the turbulence models, which renders this simulation very challenging.

Figure 3.2. *Circular cylinder at Reynolds number* 8.4×10^6. *Instantaneous Q-criterion isosurfaces (colored with velocity modulus). For a color version of this figure, see www.iste.co.uk/dervieux/meshadaptation2*

The mesh used in this simulation contains 4.3 million nodes and 25 million tetrahedra. The smallest cell thickness is 2.5×10^{-6}. The computational domain is decomposed into 192 subdomains. When integer K, used for the definition of the inner and outer zones, is set to 5, 10 and 20, the percentage of nodes located in the inner zone is 15%, 19% and 24%, respectively (see Table 3.1).

K	$nproc$	$\frac{N^{small}}{N}$ (%)	Expected gain (theoretical)	Measured gain (UP/MCP/R)	Error (%)
20	192	24	3.45	1.18/1.43/2.27	2.6×10^{-3}
60	192	27	3.48	1.21/1.52/2.32	5×10^{-3}
BDF2 CFL=30	192			20./ $-$ / $-$	1.0
CFL=2(est.)	192			1.5/ $-$ / $-$	5×10^{-3}

Table 3.1. *Circular cylinder at Reynolds number* 8.4×10^6. *Time step factor K, number of processors, percentage of nodes in the inner region, theoretical gain in scalar mode, measured parallel gain, and relative error with respect to explicit time advancing, for MR ($K = 2, 60$) and implicit BDF2 ($CFL = 30, 2$). UP: usual partition; MCP: Metis multi-constrained partition; R: analytic radial optimal partition*

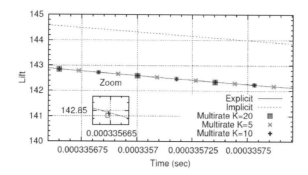

Figure 3.3. *Circular cylinder at Reynolds number* 8.4×10^6. *Zoom of the lift curves obtained with explicit, implicit and MR schemes. For a color version of this figure, see www.iste.co.uk/dervieux/meshadaptation2*

For each simulation, 192 cores were used on a Bullx B720 cluster, and the CFL number was set to 0.5. CPU times for the explicit and MR schemes with different values of K are given in Table 3.1. One can observe that the efficiency of the MR approach is rather moderate, with however a noticeable improvement of the gain in the case of a radial optimal partition. The cost of the correction step is indeed relatively high compared to the prediction step. This is certainly due to an important number of inner nodes (which implies also a moderate theoretical scalar gain) and a non-uniform distribution of these nodes among the computational cores for the usual partition.

It is interesting to compare these performances with the efficiency of an implicit algorithm. The implicit algorithm (BDF2) that we use is advanced with a GMRES linear solver using a Restrictive-Additive Schwarz preconditioner and ILU(0) in each partition (see Koobus et al. 2011 for further details). In the cases computed with the implicit scheme, the CFL is fixed to 30 and the total number of GMRES iterations for one time step is around 20. For this CFL, the gain of an implicit computation with respect to an explicit one at CFL 0.5 is measured between 12 and 22 depending on the number of nodes per processor. The BDF2 algorithm is second-order accurate in time and we shall use this property when estimating which time step reduction is necessary for reducing by a given factor the deviation with respect to explicit time advancing. An important gain is observed as compared to the MR case, but at the cost of a degradation of the accuracy (see Table 3.1 and Figure 3.3). This is probably related to the small-scale fluctuations, which arise at this Reynolds number (Figure 3.2). They need to be captured with a rather accurate time advancing. A 1% deviation after a shedding cycle may become 20% after 20 cycles and deteriorate the prediction of bulk fluctuations. In order to obtain the same level of error, the implicit time advancing, which is second-order accurate in time, should be run with a CFL of 2, with a gain of only 1.5 (less than the MR case with the radial optimal partition).

3.4.2. *Mesh adaption for a contact discontinuity*

This example is a simplified case of mesh adaptation. We consider the case of a moving contact discontinuity. For this purpose, the compressible Euler equations are solved in a rectangular parallelepiped as computational domain where the density is initially discontinuous at its middle (see Figure 3.4) while velocity and pressure are uniform.

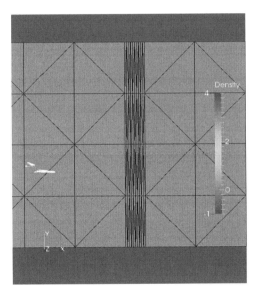

Figure 3.4. *Mesh adaptative calculation of a traveling contact discontinuity. Instantaneous mesh with mesh concentration in the middle of zoom and corresponding advected discontinuous fluid density. Density is 3 on the left, 1 on the right. For a color version of this figure, see www.iste.co.uk/dervieux/meshadaptation2*

The uniform velocity is a purely horizontal one. As can be seen in Figure 3.4, small cells are present on either side of the discontinuity. The mesh adapts analytically during the computation in such a way that the nodes located at the discontinuity are still the same, and the number of small cells are equally balanced on either side of the discontinuity. An Arbitrary Lagrangian-Eulerian formulation is then used to solve the Euler equations on the resulting deforming mesh. Our long term objective is to combine the MR time advancing with a mesh adaptation algorithm in such a way that the small time steps imposed by the necessary good resolution of the discontinuity remain of weak impact on the global computational time. The 3D mesh used in this simulation contains 25,000 vertices. The computational domain is divided into two subdomains, the partition interface being defined in such a way that it follows the center plan of the discontinuity. When integer K, used for the definition of the inner and outer zones, is set to 5, 10 and 15, the percentage of nodes located in the inner zone is always 1.3%, which corresponds to the vertices of the small cells located on either side of the discontinuity. The CFL with respect to propagation is 0.5. The MR scheme with the aforementioned values of K is used for the computation. Each simulation is run on 2 cores of a Bullx B720 cluster. In Table 3.2, CPU times (prediction step/correction step) are given for the MR approach and different time step factors K. The correction step, consists of explicit time advancing on inner zone, 1.3% of the mesh (solely 78 vertices on each partition), but, due to parallel and vector inefficiency, one Shu step of it is 39% of one Shu explicit step on the whole mesh.

K	$N^{small}(K)/N$ (%)	Expected gain (theoretical)	CPU pred. step (s/KΔt)	CPU correc. step (s/KΔt)	Measured gain (parallel)
5	1.3	4.7	0.124	0.244	1.7
10	1.3	8.8	0.124	0.482	2.0
15	1.3	12.5	0.124	0.729	2.2

Table 3.2. *Mesh adaptative propagation of a contact discontinuity: Time step factor K, CPU of the explicit scheme per explicit time-step Δt and per node, percentage of nodes in the inner region, theoretical gain in scalar mode, CPU of the prediction step per time-step $K\Delta t$, CPU of the correction step per time-step $K\Delta t$ and measured parallel gain*

3.5. Conclusion

The motivation for using a multi-rate approach is twofold. *First*, with the arising of novel anisotropic mesh adaptation methods, the complexity of unsteady accurate computations with large and small mesh sizes needs to be reduced. *Second*, we are interested by increasing the efficiency of accurate unsteady simulations. Of course, the very high Reynolds number hybrid simulations can be computed with implicit time advancing for maintaining a reasonable cpu. But in many cases this is done with an important degradation of the accuracy with respect to smaller time steps on the same mesh.

The proposed method is based on control volume agglomeration, and relies on:

– a prediction step where large time steps are used and where the fluxes for the smaller elements are evaluated on macro-cells for stability purpose;

– a correction step in which only the smaller elements of the so-called inner zone are advanced in time with a small time step.

An important interest of the method is that the modification effort in an existing explicit unstructured code is very low. Results show that the proposed MR strategy can be applied to complex unsteady CFD problems such as the prediction of three-dimensional flows around bluff bodies with a hybrid RANS/LES turbulence model. A simplified mesh adaptive calculation of a moving shock is also performed, as a preliminary test for mesh adaptation. All the numerical experiments are in parallel computed with MPI. This allows to identify the main difficulty in obtaining high computational gain, which is related with the parallel efficiency of the computations restricted to the inner zone. Because of the use of an explicit Runge–Kutta time advancing, the time accuracy of the MR scheme remains high and the dissipation remains low, as compared with an implicit computation. Only very small time scales are lost with respect to a pure explicit computation. Implicit accuracy is limited not only by the intrinsic scheme accuracy but also by the conditions required to achieve greater efficiency, which involve a sufficiently large time step and a short, parameter dependant, convergence of the linear solver performed in the time-advancing step. In contrast, explicit and MR computations are parameter safe, and the accuracy of the MR method is optimal in regions complementary to the inner zone.

3.6. Notes

Most contents of this chapter are borrowed from (Itam et al. 2019), which present some other numerical experiments.

About multi-rate works. The multi-rate methods starts from Skelboes's pioneering work using backward differentiation formulas (BDF) (Skelboe 1989) and continues to recent works dealing with hyperbolic conservation laws (Constantinescu and Sandu 2007; Sandu and Constantinescu 2009; Mugg 2012; Seny et al. 2014).

For the solution of ODEs or EDPs, explicit integration schemes are still often used because of the accuracy they can provide and their simplicity of implementation. Nevertheless, these schemes can prove to be very expensive in some situations, for example, stiff ODEs whose solution components exhibit different time scales, system of non-stiff ODEs characterized by different activity levels (fast/slow), or EDPs discretized on computational grids with very small elements. In order to overcome this efficiency problem, different strategies were developed, first in the field of ODEs, in order to propose an interesting alternative:

– multi-method schemes: for systems of ODEs containing both non-stiff and stiff parts, an explicit scheme is used for the non-stiff subsystem and an implicit method for the stiff one (Hofer 1976; Rentrop 1985; Weiner et al. 1993);

– multi-order schemes: for non-stiff system of ODEs, the same explicit method and step size are used, but the order of the method is selected according to the activity level (fast/slow) of the considered subsystem of ODEs (Engstler and Lubich 1997a);

– multi-rate schemes: for stiff and non-stiff problems, the same explicit or implicit method with the same order is applied to all subsystems, but the step size is chosen according to the activity level. The first multi-rate time integration algorithm goes back to the work of Rice (1960).

In what follows, we focus our review on the multi-rate approach. The application of such schemes was first limited to ODEs (Rice 1960; Andrus 1979; Gear 1971; Skelboe 1989; Sand and Skelboe 1992; Andrus 1993; Günther and Rentrop 1993; Engstler and Lubich 1997b,a; Günther et al. 1998, 2001; Savcenco et al. 2007) and restricted to a low number of industrial problems. In the past 15 years, the development and application of such methods to the time integration of PDEs was also performed. In particular, a few works were conducted on the system of ODEs that arise after semidiscretization of hyperbolic conservation laws (Constantinescu and Sandu 2007; Sandu and Constantinescu 2009; Mugg 2012; Kirby 2002; Seny et al. 2014; Löhner et al. 1984), and rare applications were performed in computational fluid dynamics (CFD) (Seny et al. 2014; Löhner et al. 1984) for which we are interested.

The work of Skelboe on multi-rate BDF methods, 1989 (Skelboe 1989). In this work, concerning multi-rate BDF schemes, first-order ODEs, made of a fast subsystem and a slow subsystem, are considered. In the proposed multi-rate strategy, the fast subsystem is integrated by a k-step BDF formula (BDF-k) with step length h, and the slow subsystem is integrated by the same BDF-k formula but with time step $H = qh$, where q is an integer multiplying factor. Interpolation and extrapolation values (following a Newton type formula) of the solution are used in the proposed algorithms.

As for the application part, a 2×2 test problem is considered for investigating the stability properties of the multi-rate algorithms. From this application, it appears that the proposed multi-rate algorithms are not necessarily A-stable, limiting the use of such methods.

The work of Günther and Rentrop on multi-rate Rosenbrock-Wanner (ROW) methods, 1993 (Günther and Rentrop 1993). In this work, regarding multi-rate ROW algorithms, autonomous first-order ODEs, which can be split into active and latent components, are considered. The multi-rate strategy is based on a ROW method in which a large time step H is used for the latent subsystem and a small time step $h = H/m$ is employed for advancing the active components of the solution. In the proposed multi-rate method, the latent and active components of the solution are extrapolated using a Padé approximation.

The application part concerns the simulation of electric circuits (inverter chain) leading to the solution of stiff ODEs (system of 250–4000 differential equations). A multi-rate four-step ROW method was implemented, leading to a A-stable algorithm, and a speedup up to 2.8 compared to the classical explicit four-stage Runge Kutta (RK) method.

The work of Löhner-Morgan-Zienkiewicz on explicit multi-rate schemes for hyperbolic problems with CFD applications, 1984 (Löhner et al. 1984). To our knowledge, this work is the first one on multi-rate methods that deals with applications in CFD. In the model problem, two subregions with different grid resolution are considered and the solution is advanced in time using a given explicit scheme. A large time step Δt_1 is used in the coarse subregion Ω_1, and a small time step $\Delta t_2 = \Delta t_1/n$ is employed in the fine subregion Ω_2. In the proposed multi-rate strategy, some grid points belonging to Ω_1 are added to the fine subregion Ω_2 (which becomes Ω'_2), and appropriate boundary conditions are used for both subregions Ω_1 and Ω'_2, together with mean values at some points belonging to Ω_1 and Ω'_2. The proposed multi-rate scheme was implemented with a second-order explicit finite element scheme (Taylor-Galerkin method of Donea).

Several CFD test cases were considered, which illustrate that shocks can be handled by the multi-rate method and that a speedup of 2 between the multi-rate scheme and its single-rate counterpart can be obtained.

The work of Kirby on a multi-rate forward Euler scheme for hyperbolic conservation laws, 2002 (Kirby 2002). This theoretical work presents and analyzes a multi-rate method for the solution of one-dimensional hyberbolic conservation laws. After semi-discretization by a finite volume MUSCL approach, a system of ODEs is obtained which is partitioned in fast and slow subsystems. A rather simple multi-rate scheme based on forward Euler steps is proposed for the solution of these subsystems. The fast subsystem is advanced with a small time step $\Delta t/m$, while a large time step Δt is used for the slow subsystem. No extrapolation/interpolation are performed in the proposed multi-rate strategy. It is shown that the multi-rate scheme satisfies the TVD property and a maximum principle under local CFL conditions, but it is only first order time accurate.

The work of Constantinescu and Sandu on multi-rate RK methods for hyperbolic conservation laws, 2007 (Constantinescu and Sandu 2007). One-dimensional scalar hyperbolic equations are considered in this study. For the solution of these equations, a second-order accurate multi-rate scheme that inherits stability properties of the single rate integrator (maximum principle, TVD, TVB, monotocity-preservation, positivity) is developed. After a semi-discrete finite volume approximation (which satisfies some of the above stability properties), a system of ODEs is obtained and divided into slow and fast subsystems. The computational domain is split into a subdomain Ω_F corresponding to a fast characteristic time where a small time step $\Delta t/m$ is used in the multi-rate scheme and a subdomain Ω_S with a slow characteristic time where a large time step Δt is employed. Furthermore, a buffer zone between Ω_F and Ω_S is introduced in order to bridge the transition between these two subdomains for the purpose that the multi-rate scheme satisfies the stability properties of the single rate scheme. In the proposed multi-rate method, appropriate explicit RK schemes are also used in each subdomain and the buffer zone.

The resulting multi-rate algorithm is assessed on 1D problems (advection equation, and Burger's equation) with fixed and moving grids. It was checked that the numerical solutions are second-order accurate, positive, obey the maximum principle, TVD, wiggle free, and the integration is conservative. Speedups up to 2.4 were obtained.

The work of Seny et al. on the parallel implementation of multi-rate RK methods, 2014 (Seny et al. 2014). This work focuses on the efficient parallel implementation of explicit multi-rate RK schemes in the framework of discontinuous Galerkin methods. The multi-rate RK scheme used is the approach proposed by Constantinescu and Sandu (2007) and introduced in the previous section.

In order to optimize the parallel efficiency of the multi-rate scheme, they propose a solution based on multi-constraint mesh partitioning. The objective is to ensure that the workload, for each stage of the multi-rate algorithm, is almost equally shared by each computer core, that is, the same number of elements are active on each core, while minimizing inter-processor communications. The METIS software is used for the mesh decomposition, and the parallel programing is performed with the message passing interface.

The efficiency of the parallel multi-rate strategy is evaluated on two- and three-dimensional CFD problems. It is shown that the multi-constraint partitioning strategy increases the efficiency of the parallel multi-rate scheme compared to the classical single-constraint partitioning. However, they observe that strong scalability is achieved with more difficulty with the multi-rate algorithm than with its single rate counterpart, especially when the number of processors becomes important compared to the number of mesh elements. The possible low number of elements per multi-rate group and per processor is a limiting factor for the proposed approach.

4

Goal-Oriented Adaptation for Inviscid Steady Flows

The purpose of this chapter is to introduce a mesh adaptation method, which will be closer to the objective of reducing the approximation error, taking into account the discretization of the PDE. In the *goal-oriented approach*, the user specifies *a particular scalar output* of the computation. The error in approximating this output through the discretization of the PDE is minimized. The goal-oriented approach relies on (1) a residual/local error estimate in terms of the metric, and (2) an adjoint state relating this estimate to the functional. The use of a continuous metric parameterization permits to build a smooth optimization problem solved via an optimality system. This method is an important method for this book. It is first explained for the case of steady Euler flows with application to sonic boom calculations.

4.1. Introduction

In the previous chapters, the adaptive specification of the mesh is mainly deduced from an interpolation error estimate for some fields (the sensors or the features) related to the PDE solution. Focusing on these interpolation errors is a limitation of this study. If for many applications, the simplicity of this standpoint is an advantage, there are also many applications where *feature-based adaptation* is far from being optimal regarding the way the degrees of freedom are distributed in the computational domain. Indeed, minimizing the interpolation error is not often so close to the actual objective that consists of obtaining the most accurate approximate solution of a PDE. This is particularly true in the many engineering applications where a *specific scalar output* needs to be accurately evaluated, for example, lift, drag and moment in aeronautics. Focusing on the best accuracy of such a scalar output is the purpose of *goal-oriented mesh adaptation* methods, which will take into account both the solution and the PDE in the error estimation, through error estimate and adjoint state. The formulation of goal-oriented mesh adaptation (Becker et al. 1999; Braack and Rannacher 1999; Becker and Rannacher 1996; Giles 1997a; Giles and Pierce 1999; Giles and Suli 2002a; Pierce and Giles 2000; Venditti and Darmofal 2002, 2003) has brought many improvements in the formulation and the resolution of mesh adaptation for PDE approximations. Let us write the continuous PDE and the discrete one as follows:

$$\Psi(w) = 0 \quad \text{and} \quad \Psi_h(w_h) = 0. \qquad [4.1]$$

We focus on deriving the best mesh to approximate a given functional j depending on the solution w. In other words, we examine how *to minimize the approximation error δj committed in the evaluation of the functional j*:

$$\delta j = |j(w) - j(w_h)|. \qquad [4.2]$$

This objective must be distinguished from deriving the optimal mesh to minimize a global approximation error $\|w - w_h\|$. See Becker and Rannacher (1996); Verfürth (1996) for *a posteriori* error estimates devoted to this latter task and Chapter 6 for an a posteriori standpoint. To simplify our goal-oriented formulation, we assume that the functional j is enough regular to be replaced by its Jacobian g. We simplify it as follows[1]:

$$j(w) = (g, w).$$

We also assume for simplicity that there is no discrete error evaluation on j; this means that $j_h(w_h) = j(w_h)$ [2]. On this basis, we seek for the mesh $\mathcal{H}$ that gives the smallest error for the evaluation of j from the solution w_h:

$$\min_{\mathcal{H}} |(g, w_h) - (g, w)|, \qquad [4.3]$$

where w and w_h verify state equations [4.1]. With statement [4.3], the mesh adaptation problem is recast to an optimization problem. In order to go one step forward in the analysis, we need to implicitly take into account constraint [4.1] in problem [4.3]. The

---

[1] In the general case j, the derivative of j with respect to w will be taken in adjoint RHS in place of g.
[2] Otherwise, an extra term in the error to minimize will take into account the dependency of j with respect to the mesh (i.e. to the metric).

initial approximation error on the cost functional $|(g, w_h) - (g, w)|$ can be simplified in terms of a *local error* $\Psi(w_h)$ because of the introduction of the *adjoint state* w^*:

$$(g, w_h - w) \approx \left(g, \left(\frac{\partial \Psi}{\partial w}\right)^{-1} \Psi(w_h)\right) = (w^*, \Psi(w_h)),$$

where w^* is the solution of

$$\left(\left(\frac{\partial \Psi}{\partial w}\right)^* w^*, \psi\right) = (g, \psi).$$

In practice, the exact adjoint w^* is not available. By introducing an approximate adjoint w_h^*, we get

$$(g, w_h - w) \approx (w_h^*, \Psi(w_h)).$$

4.1.1. *What to do with this estimate?*

The right-hand side is a spatial integral, the integrand of which should be used to decide where to refine the mesh. The iso-distribution of the error can be approximated by refining according to a tolerance, as in Becker and Rannacher (1996). In Giles and Suli (2002a), it is proposed to use this right-hand side as a correction that importantly improves the quality (in particular the convergence order) of the approximation of j by setting

$$j^{corrected} = (g, w_h) + (w_h^*, \Psi(w_h)).$$

However, by substituting w^* by w_h^*, we introduce an error in $O(w_h^* - w^*)$, which results in being the main error term when we use $j^{corrected}$. In Venditti and Darmofal (2002, 2003), it is proposed to keep the corrector and to adapt the mesh to this higher order error term, that is,

$$j^{corrected} - j \approx (w_h^* - w^*, \Psi(w_h)) \qquad [4.4]$$

or, equivalently,

$$j^{corrected} - j \approx \left(w - w_h, g - \left(\frac{\partial \Psi}{\partial w}\right)^* w_h^*\right). \qquad [4.5]$$

In order to evaluate numerically these terms, Venditti and Darmofal (2002, 2003) chose to approach these approximation errors, that is, $w_h^* - w^*$ and $w_h - w$, by interpolation errors:

$$w_h - w \approx L_{h/2}^h w_h - Q_{h/2}^h w_h, \qquad [4.6]$$

relying on the differences between the linear representation $L_{h/2}^h$ and a quadratic representation $Q_{h/2}^h$ reconstructed on a finer mesh.

4.1.2. *Adjoint-L^1 approach*

In the approach described in this chapter, metric analysis and goal-oriented analysis are complementary. Indeed, a metric-based method specifies the *object* of our search through an accurate description of the ideal optimal mesh, while a goal-oriented method specifies precisely the *purpose* of the search, specifying *which error* will be reduced. It is then very motivating to seek for a combination of both methods, with the hope of obtaining *a metric-based specification of the best mesh for reducing the error committed on a target functional*. A few works address this purpose. In Venditti and Darmofal (2003), an anisotropic step relying on the Hessian of the Mach number is introduced into the *a posteriori* estimate. In Rogé and Martin (2008), an ad hoc formula gives a better impact to the anisotropic component. This chapter presents a different approach to the combination of Hessian and estimates.

The first key point is to work in a continuous (non-discrete) formulation by following the continuous interpolation analysis described in Chapters 3 and 4 of Volume 1. But feature-based metric methods minimize an interpolation error, the deviation between an exact sensor and its linear interpolation on the mesh. This assumes the knowledge of the solution; this is an a priori standpoint. In contrast, goal-oriented methods are generally envisaged from an *a posteriori* standpoint; we refer to Becker and Rannacher (1996), Verfürth (1996), Apel (1999), Giles and Suli (2002a), Picasso (2003), Formaggia et al. (2004). With that option, the error committed is element-wise known on the mesh under study. Mesh refinement scheme based on a such a posteriori estimations relies on an equi-distribution principle and is thus intrinsically isotropic. Fortunately, goal-oriented methods do not need to be systematically associated with an a posteriori analysis. According to Babuška and Strouboulis (2001), a priori analysis can bring many useful information. Check these information such as anisotropy (see Formaggia and Perotto 2001). This chapter will demonstrate that the goal-oriented error can be globally expressed by an a priori analysis containing anisotropy information. For exploiting this information, a third key point is to work with a numerical scheme that allows expressing the difference $\Psi_h(w) - \Psi(w)$ in term of interpolation errors. This can be done in a straightforward way by considering finite-element variational formulations.

4.1.3. *Outline*

This chapter begins with the introduction of a theoretical abstract framework in section 4.2. Within this framework, a first a priori goal-oriented error estimate (equation [4.10]) is derived. Its application to the compressible Euler equations is then studied in section 4.3 for a class of specific Galerkin-equivalent numerical schemes. From this study, a generic anisotropic error estimate (equation [4.13]) is expressed. The estimate is then minimized globally on the abstract space of continuous meshes (section 4.4). We give in section 4.5 some details on the main modifications of the adaptive loop as compared to classical Hessian-based mesh adaptation. The practical optimal metric field minimizing the goal-oriented error estimate is then exhibited. 3D detailed examples conclude this chapter by providing a numerical validation of the theory (section 4.6).

4.2. A more accurate nonlinear error analysis

An accurate error analysis cannot be done without specifying the operator which permits to pass from continuous to discrete and vice versa. Since the P^1 interpolate Π_h is the pivot of today's metric analysis, this operator is naturally used in our analysis.

4.2.1. *Assumptions and definitions*

Let V be a Hilbert space of functions. We write the state equation under a variational statement:

$$w \in V, \ \forall \varphi \in V, \ (\Psi(w), \varphi) = 0, \qquad [4.7]$$

where the operator $(,)$ is the Hilbertian product, and w is the solution of this equation. Symbol $(\Psi(w), \varphi)$ holds for a functional that is linear with respect to test function φ but *a priori* nonlinear with respect to w. The continuous adjoint w^* is solution of

$$w^* \in V, \ \forall \psi \in V, \ \left(\frac{\partial \Psi}{\partial w}(w)\psi, w^*\right) = (g, \psi), \qquad [4.8]$$

where g is the jacobian of a given functional j. Let V_h be a subspace of $\mathcal{V} = V \cap \mathcal{C}^0$ of finite dimension N, and we write the discrete state equation as follows:

$$w_h \in V_h, \ \forall \varphi_h \in V_h, \ (\Psi_h(w_h), \varphi_h) = 0.$$

Then, we can write

$$(\Psi_h(w), \varphi_h) - (\Psi_h(w_h), \varphi_h) = (\Psi_h(w), \varphi_h) - (\Psi(w), \varphi_h)$$
$$= ((\Psi_h - \Psi)(w), \varphi_h). \quad [4.9]$$

For the a priori analysis, we assume that the solutions w and w^* are sufficiently regular to be interpolated:

$$w \in V \cap \mathcal{C}^0, \; w^* \in V \cap \mathcal{C}^0,$$

and that we have an interpolation operator:

$$\Pi_h : V \cap \mathcal{C}^0 \to V_h.$$

4.2.2. *A priori estimation*

We start from a functional defined as: $j(w) = (g, w)$, where g is a function of V. Our objective is to estimate the following approximation error on the functional:

$$\delta j = j(w) - j(w_h),$$

as a function of continuous solutions, continuous residuals and discrete residuals. The error δj is split as follows:

$$\delta j = j(w) - j(w_h) = (g, w - \Pi_h w) + (g, \Pi_h w - w_h).$$

δj is now composed of an *interpolation error* and an *implicit error*, $\Pi_h w - w_h$. Let us introduce the following discrete adjoint system:

$$w_h^* \in V_h, \; \forall \psi_h \in V_h, \; \left(\frac{\partial \Psi_h}{\partial w}(\Pi_h w)\psi_h, w_h^* \right) = (g, \psi_h).$$

We derive the following extension of δj with the choice $\psi_h = \Pi_h w - w_h$:

$$\delta j = (g, w - \Pi_h w) + \left(\frac{\partial \Psi_h}{\partial w}(\Pi_h w)(\Pi_h w - w_h), w_h^* \right).$$

According to [4.9], we have

$$(\Psi_h(\Pi_h w), w_h^*) - (\Psi_h(w_h), w_h^*) = (\Psi_h(\Pi_h w), w_h^*) - (\Psi_h(w), w_h^*) \\ + ((\Psi_h - \Psi)(w), w_h^*),$$

which gives, by using a Taylor extension,

$$\left(\frac{\partial \Psi_h}{\partial w}(\Pi_h w)(\Pi_h w - w_h), w_h^*\right) = (\Psi_h(\Pi_h w), w_h^*) - (\Psi_h(w), w_h^*) + \\ ((\Psi_h - \Psi)(w), w_h^*) + R_1,$$

where the remainder R_1 is

$$R_1 = \left(\frac{\partial \Psi_h}{\partial w}(\Pi_h w)(\Pi_h w - w_h), w_h^*\right) - (\Psi_h(\Pi_h w), w_h^*) + (\Psi_h(w_h), w_h^*).$$

Thus, we get the following expression of δj:

$$\delta j = (g, w - \Pi_h w) + (\Psi_h(\Pi_h w), w_h^*) - (\Psi_h(w), w_h^*) + ((\Psi_h - \Psi)(w), w_h^*) + R_1,$$

We now apply a second Taylor extension to get

$$(\Psi_h(\Pi_h w), w_h^*) - (\Psi_h(w), w_h^*) = \left(\frac{\partial \Psi_h}{\partial w}(w)(\Pi_h w - w), w_h^*\right) + R_2,$$

with remainder term

$$R_2 = (\Psi_h(\Pi_h w), w_h^*) - (\Psi_h(w), w_h^*) - \left(\frac{\partial \Psi_h}{\partial w}(w)(\Pi_h w - w), w_h^*\right).$$

This implies

$$\delta j = (g, w - \Pi_h w) + \left(\frac{\partial \Psi_h}{\partial w}(w)(\Pi_h w - w), w_h^*\right) \\ + ((\Psi_h - \Psi)(w), w_h^*) + R_1 + R_2.$$

In contrast to an a posteriori analysis, this analysis starts with a discrete adjoint w_h^*. However, our purpose is to derive a continuous description of the main error term.

Thus, we get rid of the discrete solutions in the dominating terms. To this end, we re-write δj as follows:

$$\delta j = (g, w - \Pi_h w) + \left(\frac{\partial \Psi}{\partial w}(w)(\Pi_h w - w), w^*\right)$$
$$+ ((\Psi_h - \Psi)(w), w^*) + R_1 + R_2 + D_1 + D_2 + D_3,$$

where

$$D_1 = \left(\left(\frac{\partial \Psi_h}{\partial w} - \frac{\partial \Psi}{\partial w}\right)(w)(\Pi_h w - w), w_h^*\right),$$
$$D_2 = \left(\frac{\partial \Psi}{\partial w}(w)(\Pi_h w - w), w_h^* - w^*\right),$$
$$D_3 = ((\Psi_h - \Psi)(w), w_h^* - w^*).$$

The latter expression of δj can be even more simplified because of the continuous adjoint of equation [4.8], leading to

$$\delta j = ((\Psi_h - \Psi)(w), w^*) + R_1 + R_2 + D_1 + D_2 + D_3. \qquad [4.10]$$

At least formally, the R_i and the D_k are higher order terms, and the first term in the right-hand side of [4.10] is the dominating one. In order to exhibit, from [4.10], a formulation specifying the optimal mesh, we need to make the physical and numerical models more precise.

4.3. The case of the steady Euler equations

In this section, we study how equation [4.10] can be transformed in the context of the steady Euler equations. We restrict to the particular discretization of Chapter 1 of Volume 1.

4.3.1. *Variational analysis*

Equation [1.5] from Volume 1 will play the role of equation [4.7] in the abstract analysis of the previous section. For the discretisation, we consider a discrete domain Ω_h and a discrete boundary Γ_h, which are not necessarily identical to the continuous ones. The MUSCL scheme introduced in Chapter 1 of Volume 1 can be written as a variational Galerkin-type scheme complemented by a higher order term, which can be assimilated to a numerical viscosity term D_h. This gives, $\forall \phi_h \in V_h$,

$$\int_{\Omega_h} \phi_h \nabla . \mathcal{F}_h(W_h) \, d\Omega_h - \int_{\Gamma_h} \phi_h \hat{\mathcal{F}}_h(W_h) . \mathbf{n} \, d\Gamma_h = - \int_{\Omega_h} \phi_h \, D_h(W_h) d\Omega_h. \quad [4.11]$$

The numerical diffusion term is of higher order as soon as it is applied to the interpolation of a smooth enough field W on a sufficiently regular mesh (see Mer 1998 for a detailed analysis):

$$\left| \int_{\Omega_h} \phi_h \, D_h(W_h) d\Omega_h \right| \leq h^3 K(W) |\phi_h|_{L^2}.$$

As a result, the dissipation term will be neglected in the same way we neglect the remainders R_i and D_k of relation [4.10]. Even in the case of a flow with shocks, we get inspired by the Hessian-based study of previous chapters and choose to neglect the error term from artificial dissipation.

4.3.2. *Approximation error estimation*

Returning to the output functional $j(W) = (g, W)$ and according to estimate [4.10], the main term of the *a priori* error estimation of δj becomes

$$\delta j = (g, W - W_h) \approx ((\Psi_h - \Psi)(W), W^*) \quad \text{with} \quad \frac{\partial \Psi}{\partial W} W^* = g,$$

where W^* is the continuous adjoint state. Using the exact solution W in equations [1.5] and [4.11] while neglecting the dissipation D_h leads to

$$(g, W - W_h) \approx \int_{\Omega_h} W^* \left(\nabla . \mathcal{F}_h(W) - \nabla . \mathcal{F}(W) \right) d\Omega_h$$

$$- \int_{\Gamma_h} W^* \left(\hat{\mathcal{F}}_h(W) - \hat{\mathcal{F}}(W) \right) . \mathbf{n} \, d\Gamma_h.$$

By integrating by parts the previous estimate, we get

$$(g, W - W_h) \approx \int_{\Omega_h} \nabla W^* \left(\mathcal{F}(W) - \mathcal{F}_h(W) \right) d\Omega_h$$

$$- \int_{\Gamma_h} W^* \left(\bar{\mathcal{F}}(W) - \bar{\mathcal{F}}_h(W) \right) . \mathbf{n} \, d\Gamma_h,$$

where fluxes $\bar{\mathcal{F}}$ are given by

$$\bar{\mathcal{F}}(W).\mathbf{n} = \mathcal{F}(W).\mathbf{n} - \hat{\mathcal{F}}(W).\mathbf{n}.$$

By definition, $\mathcal{F}_h$ is the linear interpolate of $\mathcal{F}$, that is, $\Pi_h \mathcal{F} = \mathcal{F}_h$, thus we have

$$\delta j \approx \int_{\Omega_h} \nabla W^* \left(\mathcal{F}(W) - \Pi_h \mathcal{F}(W)\right) d\Omega_h - \int_{\Gamma_h} W^* \left(\bar{\mathcal{F}}(W) - \Pi_h \bar{\mathcal{F}}(W))\right).\mathbf{n} \, d\Gamma_h.$$

[4.12]

We observe that this estimate of δj is expressed in terms of interpolation errors for the fluxes and in terms of the continuous functions W and W^*. The integrands in [4.12] contain positive and negative parts, which could compensate for some particular meshes. In our strategy, we prefer to avoid these parasitic effects in our estimate[3]. To this end, all integrands are bounded by their absolute values:

$$(g, W_h - W) \leq \int_{\Omega_h} |\nabla W^*||\mathcal{F}(W) - \Pi_h \mathcal{F}(W)| \, d\Omega_h$$
$$+ \int_{\Gamma_h} |W^*||(\bar{\mathcal{F}}(W) - \Pi_h \bar{\mathcal{F}}(W)).\mathbf{n}| d\Gamma_h.$$

[4.13]

4.4. Error model minimization

Starting from the bound in [4.13], several options are possible to derive an optimal mesh for the observed functional. We propose to work in the continuous mesh framework by building a completely continuous estimate, which is possible with the a priori estimate standpoint. It allows us to define proper differentiable optimization (Arsigny et al. 2006; Absil et al. 2008) and use the calculus of variations that is undefined on the class of discrete meshes. This framework lies in the class of metric-based methods. Working in this framework enables us, as in the previous section, to write estimate [4.13] in a continuous form:

$$(g, W_h - W) \approx E(\mathcal{M}) = \int_{\Omega} |\nabla W^*||\mathcal{F}(W) - \pi_{\mathcal{M}} \mathcal{F}(W)| d\Omega$$
$$+ \int_{\Gamma} |W^*||(\bar{\mathcal{F}}(W) - \pi_{\mathcal{M}} \bar{\mathcal{F}}(W)).\mathbf{n}| d\Gamma,$$

[4.14]

---

3 In other words, we prefer to locally overestimate the error.

where $\mathcal{M} = (\mathcal{M}(\mathbf{x}))_{\mathbf{x} \in \Omega}$ is a continuous mesh defined by a Riemannian metric field and $\pi_{\mathcal{M}}$ is the continuous linear interpolate defined hereafter. We are now focusing on the following (continuous) mesh optimization problem:

$$\text{Find } \mathcal{M}_{opt} = \text{Argmin}_{\mathcal{M}} \, E(\mathcal{M}). \qquad [4.15]$$

A constraint is added to the previous problem in order to bound mesh fineness. In this continuous framework, we impose the metric complexity to be equal to a specified positive integer N. Let us split the functional $E(\mathcal{M})$ into a volumic functional and a surfacic one. To the minimization of both correspond an optimal metric made of a volume tensor field $\mathcal{M}_{go}$ defined in Ω and a surface optimal metric $\bar{\mathcal{M}}_{go}$ defined on Γ. This is performed as follows:

– For any $\mathbf{x}$ of Ω, the pseudo-Hessian matrix arising from the Euler fluxes is written as

$$H(\mathbf{x}) = \sum_{j=1}^{5} \left| \frac{\partial W^*}{\partial x} \right| \left| H([\mathcal{F}_1(W)]_j) \right|, + \left| \frac{\partial W^*}{\partial y} \right| \left| H([\mathcal{F}_2(W)]_j) \right|,$$

$$+ \left| \frac{\partial W^*}{\partial z} \right| \left| H([\mathcal{F}_3(W)]_j) \right|,$$

with $[\mathcal{F}_i(W)]_j$ denoting the jth component of the vector $\mathcal{F}_i(W)$.

– For any $\mathbf{x}$ of Γ, the pseudo-Hessian matrix arising from the surface contribution is written as

$$\bar{H}(\mathbf{x}) = \sum_{j=1}^{5} \left| W^* \right| \left| H\left(\sum_{i=1}^{3} \bar{\mathcal{F}}_i(W) n_i \right) \right|,$$

where $\mathbf{n} = (n_1, n_2, n_3)$ is the outward normal of Γ.

We can separately apply the optimum computation of section 4.4.1 in Chapter 3 to the two functionals

$$\mathcal{M}^{go}(\mathbf{x}) = C \, \det(|H(\mathbf{x})|)^{-\frac{1}{5}} |H(\mathbf{x})| \qquad [4.16]$$

and $\bar{\mathcal{M}}^{go}(\mathbf{x}) = \bar{C} \, \det(|\bar{H}(\mathbf{x})|)^{-\frac{1}{4}} |\bar{H}(\mathbf{x})|.$

Constants C and $\bar{C}$ depend on the desired complexity N.

4.5. Adaptative strategy

By freezing the continuous solution, we have in some manner linearized the continuous metric problem, that is, made the error *quadratic* and got an intermediate optimal metric with a close formula. The optimal metric, solution of the continuous goal-oriented mesh adaptation problem, is the fixed point of the *nonlinear coupling* between the quadratic optimum and the mapping $\mathcal{M} \mapsto W_{\mathcal{M}}$. As for most PDE-related problems and as mesh adaptation problems of previous chapters, it will be *numerically approximated*, in particular by using the mapping $\mathcal{M} \mapsto \mathcal{H}(\mathcal{M}) \mapsto W_{\mathcal{H}(\mathcal{M})}$, where $\mathcal{H}(\mathcal{M})$ is a unit mesh of $\mathcal{M}$.

The practical adaptative strategy is similar to previous anisotropic metric-based mesh adaptation. As both the solution and the mesh are changing each time, the metric is computed, a nonlinear loop is set up in order to converge toward a fixed point for the couple mesh-solution. Algorithm 4.1 is used for the adjoint-based mesh adaptation for a steady problem.

Algorithm 4.1. Adjoint-based mesh adaptation for a steady problem

Initial mesh $\mathcal{H}_0^0$, solution W_0^0, adjoint $W_0^{*,0}$, and complexity $\mathcal{C}^0$

while $i \leq n_{adap}$ **do**

    1) Compute optimal metric for the considered error estimate and complexity $\Longrightarrow \mathcal{M}_{i-1}$

    2) Generate new adapted mesh $\Longrightarrow \mathcal{H}_i$

    3) Interpolate primal and adjoint states on the new mesh $\Longrightarrow (W^0)_i$ and $(W^{*,0})_i$

    4) Compute primal state $\Longrightarrow W_i$

    5) Compute adjoint state $\Longrightarrow W_i^*$

Endwhile.

From an initial triple mesh-solution-adjoint $(\mathcal{H}_0, \mathcal{W}_0, \mathcal{W}_0^*)$ (converged on current mesh), it is composed of the following sequences. At step i, a metric tensor field $\mathcal{M}_i$ is deduced from $(\mathcal{H}_i, \mathcal{W}_i, \mathcal{W}_i^*)$ because of the anisotropic error estimate. The latter is used by the adaptive mesh generator that generates a unit mesh with respect to $\mathcal{M}_i$. The previous solution and adjoint are then linearly interpolated on the new mesh, and iteratively converged to solution and adjoint solutions on the current mesh $\mathcal{H}_i$. This procedure is repeated until convergence of the couple mesh-solution.

In next section, we investigate the novelties to be introduced when dealing with the adjoint-based anisotropic error estimate. The main modifications concern the flow and adjoint solver and the remeshing stage. In this section, the following notations are used. $\mathcal{H}$ denotes the mesh of the domain Ω_h, $\partial\mathcal{H}$ the mesh of the boundary Γ_h of Ω_h, W_h is the state provided by the flow solver and $j(W_h)$ the observed functional defined on $\gamma \subset \Omega_h$.

4.5.1. Adjoint solver

As compared to Feature-based mesh adaptation, an extra step to add to the solver is the resolution of the linear system providing the adjoint state: $A_h^* W_h^* = g_h$, where g_h is the approximated jacobian of $j(W_h)$ with respect to the conservative variables and W_h^* is the adjoint state. The state is solved with the linearized implicit pseuso-unsteady method using a spatially first-order accurate Jacobian A_h. It is described in Chapter 1 of Volume 1. We define A_h^* as the adjoint of A_h. In the Wolf code, the linear systems for W_h^* (as for W_h) is solve with the iterative GMRES method coupled with an incomplete $BILU(0)$ preconditioner (Saad 2003). Once W_h^* is computed, its vertex gradient

$$\left(\frac{\partial W_h^*}{\partial x}, \frac{\partial W_h^*}{\partial y}, \frac{\partial W_h^*}{\partial z}\right)$$

is recovered by using a $\mathbf{L}^2$ projection from the neighboring element-wise constant gradients (Alauzet and Loseille 2009a).

4.5.2. Optimal goal-oriented discrete metric

The continuous optimal metric discussed in section 4.4 is composed of a volume tensor field $\mathcal{M}_{go}$ defined in Ω and a surface one $\bar{\mathcal{M}}_{go}$ defined on Γ_h. For the discrete case, we compute the following:

– for each vertex $\mathbf{x}$ of $\mathcal{H}$, the pseudo-Hessian matrix arising from the volume contribution of each component of the Euler fluxes:

$$H(\mathbf{x}) = \sum_{j=1}^{5} ([\Delta x]_j + [\Delta y]_j + [\Delta z]_j),$$

where

$$[\Delta x]_j = \left|\frac{\partial W_h^*}{\partial x}\right| |H_R([\mathcal{F}_1(W_h)]_j)|, \quad [\Delta y]_j = \left|\frac{\partial W_h^*}{\partial y}\right| |H_R([\mathcal{F}_2(W_h)]_j)|,$$

$$[\Delta z]_j = \left|\frac{\partial W_h^*}{\partial z}\right| |H_R([\mathcal{F}_3(W_h)]_j)|,$$

with $[\mathcal{F}_i(W_h)]_j$ denoting the jth component of the vector $\mathcal{F}_i(W_h)$;

– for each vertex **x** of $\partial\mathcal{H}$, the pseudo-Hessian matrix arising from the surface contribution:

$$\bar{H}(\mathbf{x}) = \sum_{j=1}^{5} \left|W_h^*\right| \left|H_R\left(\sum_{i=1}^{3} \bar{\mathcal{F}}_i(W_h) n_i\right)\right|,$$

where $\mathbf{n} = (n_1, n_2, n_3)$ is the outward normal of Γ.

H_R stands for the operator that recovers numerically the second derivatives of an initial piecewise linear by element solution field. The recovery method is based on the Green formula method, described in section 5.4.3 of Volume 1. The standard $\mathbf{L}^1$ norm normalization is then applied independently on each metric tensor field:

$$\mathcal{M}^{go}(\mathbf{x}) = C \, \det(|H(\mathbf{x})|)^{-\frac{1}{5}} |H(\mathbf{x})| \text{ and } \bar{\mathcal{M}}^{go}(\mathbf{x}) = \bar{C} \, \det(|\bar{H}(\mathbf{x})|)^{-\frac{1}{4}} |\bar{H}(\mathbf{x})|.$$

[4.17]

Constants C and $\bar{C}$ depend on the desired complexity N.

A natural synthesis between the volumic pseudo-Hessian H and the surfacic pseudo-Hessian $\bar{H}$ is to define the final optimal metric as

$$\mathcal{M}_{opt}(\mathbf{x}) = \begin{cases} \mathcal{M}_{vol}(\mathbf{x}) & \text{for } \mathbf{x} \in \Omega \\ \mathcal{M}_{vol}(\mathbf{x}) \cap \mathcal{M}_{sur \ *-1ptf}(\mathbf{x}) & \text{for } \mathbf{x} \in \Gamma \end{cases}.$$

Of course, a special decimation[4] needs to be applied to match smoothly the boundary metric and the volumic one. In practice, numerical experiments (see Loseille 2008) show that the resulting functional prediction is not notably improved by accounting the boundary metric and that measured convergence order slightly degrades due to the extra vertices. A particular case is the case of a symmetric flow. Computing the complete flow leads to having no special boundary term on the symmetry plan, while computing the half flow adds a couple vertices near the symmetry plan. This tends to show that at least with slip conditions the boundary metric is not necessary[5].

In the presented experiment, the surfacic terms of the adaptation criterion are neglected.

---

4 Decimation: mesh size smooth transition between a fine region and a coarse region.

5 Several experiments reported in Loseille et al. (2010b) have been performed with the intersection of volumic metric and surfacic metric and compared with the use of solely the volumic metric. Up to now, no computation could indicate an advantage in using the surfacic metric.

REMARK 4.1 (Non-convergence of goal-oriented formulation).– We want to emphasize that the goal-oriented formulation is designed to favor the best convergence of the prediction of the scalar functional, but it is not designed to produce the flow field convergence. This point is illustrated in next section.

4.5.3. *Controlled mesh regeneration*

We use Yams (Frey 2001) for the adaptation of the surface and an anisotropic extension of Gamhic (George 1999) for the volume mesh. When the surface is not adapted, we use Mmg3d (Dobrzynski and Frey 2008).

4.6. Numerical outputs

4.6.1. *High-fidelity pressure prediction of an aircraft*

We consider the flow around a supersonic business jet (SSBJ). The code Wolf is used. The geometry provided by Dassault-Aviation is depicted in Figure 6.3 (left) of Volume 1. Flight conditions are Mach 1.6 with an angle of attack of $3°$. As for a body flying at a supersonic speed, each geometric singularity generates a shock wave having a cone shape; a multitude of conic shock waves are emitted by the aircraft geometry. They generally coalesce around the aircraft while propagating to the ground. The goal, here, is to compute accurately the pressure signature only on a plane located 100 m below the aircraft. The scope of this test case is to evaluate the ability of the adjoint to prescribe refinements only in areas that impact the observation region. The functional is given by

$$j(W) = \frac{1}{2} \int_\gamma \left(\frac{p - p_\infty}{p_\infty}\right)^2 \mathrm{d}\gamma,$$

with $\gamma = \{100 \leq x \leq 140,\ -1 \leq y \leq 1,\ z = -100\}$. [4.18]

In order to exemplify how adjoint-based mesh adaptation gives an optimal distribution of the degrees of freedom to evaluate the functional, this adaptation is compared to the Hessian-based mesh adaptation (the adaptation is done on the local Mach number and the interpolation error is controlled in $\mathbf{L}^2$ norm) presented in section 6.4 of Volume 1.

We apply the convergent steady fixed point Algorithm 4.1. The adaptive loop is divided into five outer steps. The complexity sequence is [10,000, 20,000, 40,000, 80,000, 160,000]. Each step is composed of six sub-iterations at constant complexity for a total of 30 adaptations. Hessian-based and adjoint-based final adapted meshes are composed of almost 800,000 vertices. They are represented in Figure 4.1 where several cuts in the final adapted meshes for the adjoint-based (top) and Hessian-based (bottom) adaptations are displayed.

For the Hessian-based adaptation, the mesh is adapted in the whole computational domain along the Mach cones and in the wake (see Figure 4.1, right). Such an anisotropic adapted mesh provides an accurate solution everywhere in the domain. But, if the aim is to only compute an accurate pressure signature on surface γ, then we clearly notice that a large amount of degrees of freedom is wasted in the upper part of the domain and in the wake where accuracy is not needed.

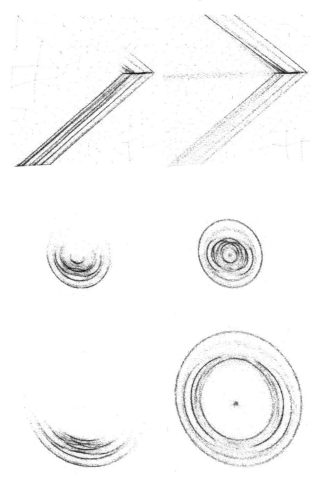

Figure 4.1. *Cut planes through the final adapted meshes for the adjoint-based (left) and Hessian-based (right) methods. Top, a cut in the symmetry plane and, middle and bottom, two cuts with an increasing distance behind the aircraft orthogonal to its path*

With regard to the adjoint-based adaptation, the mesh is mainly adapted below the aircraft in order to capture accurately all the shock waves that impact the observation plane (Figure 4.1, left). On the contrary, areas that do not impact the functional are ignored with the goal-oriented approach: the region over the aircraft, the wake of the SSBJ and in the region just behind the aircraft, only the lower half of the conic shock waves are refined and the angular amplitude of the refined part keeps on decreasing along with the distance to the aircraft.

Another point of interest, which is more technical, is that the Hessian-based adaptation in L^2 norm prescribes a mesh size that depends on the shock intensity. A stronger shock is then more refined than a weaker one. In this simulation, shocks directly below the aircraft have a lower intensity than lateral or upper shocks emitted by the wings (see Figure 4.1, bottom). Consequently, the adapted meshes with the Hessian-based method are less accurate in regions that directly affect the observation plane. On the contrary, with the adjoint-based strategy, shocks below the aircraft are not uniformly refined. The shock waves are all the more refined as they are influent on the functional independently of their amplitudes. This has a drastic consequence on the accuracy of the observed functional (Figure 4.2).

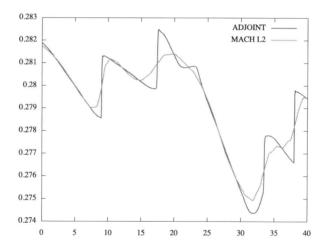

Figure 4.2. *Pressure signature along x axis in the observation plane Adjoint based calculation produces very stiff shock capturing (quasi vertical segments). For a color version of this figure, see www.iste.co.uk/dervieux/meshadaptation2*

As a conclusion, it is clear in the present computation that the Hessian-based adaptation gives a non-optimal result for the accuracy of the output functional. It gives an inappropriate distribution of the degrees of freedom. It is also demonstrated how the adjoint defines an optimal distribution of the degrees of freedom for the specific target. However, it is important to note that the mesh obtained with the Hessian-based strategy is somewhat optimal to evaluate globally the local Mach number field. It is also worth mentioning that the present method is completely automatic and gives an optimal result.

4.7. Conclusion

This chapter has described a method providing the anisotropic adapted mesh minimizing a first error term estimate in the approximation of a functional depending on the solution of a flow problem. This method is based on a formal *a priori* estimate of the functional approximation error and its minimization in an abstract continuous framework. The estimate is rather pessimistic since it does not take into account possibly compensating errors. In Chapter 5 of this volume, we discuss a more efficient estimate. With the present estimate, the goal-oriented anisotropic adaptation has been applied to the compressible Euler equations. The method has the following features:

– it produces an optimal anisotropic metric uniquely specified as the optimum of a functional and explicitly given by variational calculus from the continuous state and the adjoint state. The coupled system of the metric and the two states is the object of the discretization. This should be put in contrast with the usual process of starting from a (discrete) mesh and then improving it;

– to apply it, there is no need to choose in a more or less arbitrary way any local refinement "criterion" and no need to fix any parameter except the total number of vertices which represents the error threshold;

– mesh convergence is performed in a natural way by increasing the metric complexity at each stage of the mesh adaptation process.

The goal-oriented method is illustrated on a challenging 3D sonic boom problem. Other calculations are presented in Loseille (2008); Loseille et al. (2010a,b) and Alauzet and Loseille (2010). Numerical experiments show that the goal-oriented method enjoys at best level the advantages of Hessian-based anisotropic methods and of goal-oriented methods. As compared to the Hessian-based method, the anisotropic stretching of the meshes is not lost but even more strengthened and better distributed along shocks. Compared to traditional goal-oriented methods, the anisotropic goal-oriented method behaves like a goal-oriented method, but also naturally takes the anisotropy related to the functional into account.

4.8. Notes

This chapter revisits the study of Loseille et al. (2010b).

A paradox. O. Pironneau raised some paradox in the proposed goal-oriented adaptation algorithm. The question was: How can you adequately adapt with a criterion expressed in terms of the state and adjoint state if you do not adapt for a *better state* and a *better adjoint*?

This question is related to a particular step in the design of the mesh adaptation algorithm. Indeed, the first step in our theory is to apply a *continuous* metric analysis, which produces a continuous optimality system involving continuous state, continuous adjoint and continuous stationarity with respect to the metric.

In a second step, we have to *discretize* the whole optimality system. For this, an option *could be* to use a sequence of uniformly refined meshes. State and adjoint would be well refined. This option would finally produce (with a prohibitive computational cost) the optimal metric.

Once we have a good approximation of the optimal metric, we can generate unit adapted meshes based on it. By construction, these meshes are adequate for the evaluation of the functional through the evaluation of the state. In other words, the state computed with an unit mesh of the metric is a good mesh for the functional evaluation. At the same time, with this adapted mesh, the state is not well computed since only its features influencing the functional are supposed to be well computed. Adapting for a better state would then be *not optimal*.

Adapting for a better adjoint is a somewhat *second-order* approach, since it does not influence directly the quality of the approximation of the functional, but solely how close is the metric from the perfectly optimal one: formally, an ε-large deviation of the (unconstrained) quasi-optimum provokes an ε^2-large deviation of the (smooth) functional value.

Lastly, an examination of [4.4] and [4.5] show the adjoint itself does not need to be accurately computed, as far as RHS of [4.4] is accurately computed. Now this RHS is equal to the RHS of [4.5], which is minimized in the present method. □

5

Goal-Oriented Adaptation for Viscous Steady Flows

Chapter 4 introduces for the steady Euler flow model the different steps of a method for goal-oriented anisotropic adaptation. A continuous adjoint and a continuous error analysis are used for exhibiting an intermediate optimal continuous metric. The final optimal metric is solution of a (continuous) fixed point. A way of discretizing this fixed point is then introduced and constitutes the goal-oriented anisotropic mesh adaptation algorithm. This chapter extends the goal-oriented anisotropic adaptation to viscous steady compressible flows. Only the error analysis is different. With viscous terms, it is more complex. Further, due to the *thin boundary layers* in the flow, the efficiency of adaptation is much more influenced by the quality of the error estimate. The central principle of this kind of analysis is again to express the right-hand side of the error equation, often referred as the local error, as a function of the interpolation error of a collection of fields present in the nonlinear partial differential equations. Two approaches are considered and compared. Applications to mesh adaptive calculations of flows past airfoils and a wing-body combination are discussed.

5.1. Introduction

This chapter focuses on the building of an anisotropic goal-oriented mesh adaptation method for viscous flows. The way to derive such a method for inviscid flows was described in the previous chapter. In order to extend it to an elliptic or parabolic model, we need an adequate derivation of the error estimation when second spatial derivatives are present in the PDE. We first focus on an elliptic generic model and propose *two adjoint-based a priori analyses*. We then extend these analyses to the compressible Navier–Stokes system, which involves nonlinear parabolic terms.

For the *first a priori analysis*, we demonstrate by manipulating the nonlinear viscous terms that each of them can be written as a combination of an elliptic terms (on which the above-mentioned error estimation applies) and higher-order error terms (which can be neglected). The *second a priori analysis* is identified as satisfying some important criteria, namely a smaller set of interpolation error restricted to the conservative variables, and as giving a more compact formulation allowing for possible compensation in errors.

The chapter is outlined as follows. Section 5.2 proposes the two a priori estimates for the Poisson problem. The two goal-oriented error estimates for the Navier–Stokes equations are given in section 5.3. Finally, section 5.4 states how the optimal discrete meshes is obtained, and section 5.5 presents an application to a high-lift flow.

5.2. Case of an elliptic problem

5.2.1. *A priori finite-element analysis (first estimate)*

Standard a priori estimates have been early derived in $H^1(\Omega)$ ("projection property"), and in $L^2(\Omega)$ (Aubin–Nitsche analysis), but only by means of inequalities. The leading term of the error is generally not exhibited, but only bounds of it are proposed in term of H^2 norm of the unknown. In this section, we go a little further in order to evaluate a *first term* in the upper bound in the second-order terms. Let us focus on the Poisson problem set on domain $\Omega \subset \mathbb{R}^d$:

$$-\Delta u = f \text{ on } \Omega \text{ ; } u = 0 \text{ on } \partial\Omega. \qquad [5.1]$$

Its variational form is

$$a(u,v) = \int_\Omega \nabla u.\nabla v \, dx = (f, v), \ \forall v \in V, \qquad [5.2]$$

where V holds for the Sobolev space $V = H_0^1(\Omega) = \{u \in L^2(\Omega), \nabla u \in (L^2(\Omega))^d, u_{|\partial\Omega} = 0\}$. In order to derive an a priori estimate, we assume that solution u has some *extra smoothness*:

$$u \in \mathcal{V} = V \cap \mathcal{C}^3(\bar{\Omega}),$$

where $\mathcal{C}^3(\bar{\Omega})$ is the set of functions of class C^3 on $\Omega \cup \partial\Omega$[1]. Let $\mathcal{H}$ be a mesh of Ω made of simplices (triangles in 2D and tetrahedra in 3D): $\mathcal{H} = \bigcup_k K_k$. Let V_h be the subspace of V of continuous functions that are P^1 on each element of the mesh:

$$V_h = \left\{ \varphi_h \in V \mid \varphi_{h|K} \text{ is affine } \forall K \in \mathcal{H} \right\}.$$

The discrete variational problem is then defined by

$$a(u_h, v_h) = (f, v_h), \quad \forall\, v_h \in V_h. \tag{5.3}$$

Let Π_h be the *linear interpolation operator* Π_h defined and analysed in section 4.2 of Volume 1. In a simplified goal-oriented analysis, we are interested in estimating $(g, u_h - u)$, which can be split into two components:

$$(g, u_h - u) = (g, u_h - \Pi_h u) + (g, \Pi_h u - u), \tag{5.4}$$

where we recognize in the second difference $\Pi_h u - u$ the *interpolation error*, and the first difference $u_h - \Pi_h u$ is referred as the *implicit error*. Introducing the continuous and the discrete adjoint states u^* and u_h^* verifying

$$a(\psi, u^*) = (g, \psi), \; \forall \psi \in V \quad \text{and} \quad a(\psi_h, u_h^*) = (g, \psi_h), \; \forall \psi_h \in V_h,$$

we get

$$(g, u_h - u) = a(u_h - \Pi_h u, u_h^*) + (g, \Pi_h u - u). \tag{5.5}$$

The second term of the right-hand side can be estimated without any difficulty using corollary 4.1 of Volume 1:

$$|(g, \Pi_h u - u)| \preceq \int_\Omega |g|\,|u - \Pi_h u|\, d\Omega \preceq \int_\Omega c_1 |g|\, \text{trace}\left(\mathcal{M}^{-\frac{1}{2}} |H_u| \mathcal{M}^{-\frac{1}{2}}\right) d\Omega,$$

while it would have been a lot more complicated using the equality $(g, \Pi_h u - u) = a(\Pi_h u - u, u^*)$. Analyzing the first term of the right hand-side of relation [5.5] leads to study the following term:

$$a(u_h - \Pi_h u, \Pi_h \varphi),$$

---

[1] Functions of class C^3 are continuous together with their derivatives up to order three.

where φ is any sufficiently smooth function. To this end, we first express the implicit error term as a function of the interpolation error. It is useful to remark that the discrete statement is equivalently written as

$$a(u_h, \Pi_h\varphi) = (f, \Pi_h\varphi), \ \forall\, \varphi \in \mathcal{V}. \tag{5.6}$$

Using relation [5.6] and then relation [5.2], we get

$$a(u_h - \Pi_h u, \Pi_h\varphi) = a(u_h, \Pi_h\varphi) - a(\Pi_h u, \Pi_h\varphi)$$
$$= (f, \Pi_h\varphi) - a(\Pi_h u, \Pi_h\varphi) = a(u, \Pi_h\varphi) - a(\Pi_h u, \Pi_h\varphi), \tag{5.7}$$

which gives[2]

$$a(u_h - \Pi_h u, \Pi_h\varphi) = a(u - \Pi_h u, \Pi_h\varphi), \ \forall\, \varphi \in \mathcal{V}.$$

Then we shall use the following estimate, the proof of which can be found in Belme et al. (2019):

LEMMA 5.1.– For any couple of smooth functions (u, φ), where u is not necessarily a solution of problem [5.1], we have the following bounds:

$$\left| \int_\Omega \frac{\partial}{\partial x_i}(u - \Pi_h u) \frac{\partial}{\partial x_j} \Pi_h\varphi\, d\Omega \right| \preceq K_d \int_\Omega |\rho_H(\varphi)|\, |u - \Pi_h u|\, d\Omega + BT$$

$$\left| \int_\Omega \frac{\partial}{\partial x_i} u (u - \Pi_h u) \frac{\partial}{\partial x_j} \Pi_h\varphi\, d\Omega \right| \preceq K_d \int_\Omega |\rho_H(\varphi)|\, |u|\, |u - \Pi_h u|\, d\Omega + BT,$$

where $K_d = 3$ in two dimensions, $K_d = 6$ in three dimensions and $A \preceq B$ holds for a majoration asymptotically valid, that is, $A \leq B + o(A)$ when mesh size tends to zero. Expression $|\rho_H(\varphi)|$ holds for spectral radius of $H(\varphi)$, which is the Hessian of φ, that is, the largest (in absolute value) eigenvalue of $H(\varphi)$. The boundary terms BT are not used in the sequel. □

The estimate of lemma 5.1 is successfully tested for mesh adaptation in the sequel in Chapter 6, in more detail in Brèthes and Dervieux (2016) and also compared with another estimate in Brèthes and Dervieux (2017).

---

[2] Note that $u - \Pi_h u$ is not solution of the discrete adjoint system since u is not in V_h.

5.2.2. *Goal-oriented adaptation according to lemma 5.1*

We keep the notations introduced in Chapter 4, and in particular, we minimize the error $\delta j_{goal}(\mathcal{M})$ depending on a metric $\mathcal{M}$ and done in the evaluation of the scalar output $j = (g, u)$ by discretizing the flow under study with a mesh defined by the metric $\mathcal{M}$. The scalar error considered is simplified as follows:

$$\delta j_{goal}(\mathcal{M}) = |(g, u - u_h)| = |(g, \Pi_h u - u_h + u - \Pi_h u)|.$$

Let us define the discrete adjoint state $u_{g,h}^*$:

$$\forall \psi_h \in V_h, \quad a(\psi_h, u_{g,h}^*) = (\psi_h, g).$$

Then

$$(g, \Pi_h u - u_h + u - \Pi_h u) = a(\Pi_h u - u_h, u_{g,h}^*) + (g, u - \Pi_h u)$$

and, using [1.8],

$$(g, \Pi_h u - u_h + u - \Pi_h u) = a(\Pi_h u - u, u_{g,h}^*) + (f - \Pi_h f, u_{g,h}^*) + (g, u - \Pi_h u),$$

thus

$$\delta j_{goal}(\mathcal{M}) \approx |a(\Pi_h u - u, u_{g,h}^*) + (f - \Pi_h f, u_{g,h}^*) + (g, u - \Pi_h u)|.$$

Recall that u is unknown. The second term $\Pi_h u - u$, similar to the main term of the Hessian-based adaptation in section 6.2.1, can be explicitly approached in the same way, that is, introducing the continuous interpolation error of corollary 4.1 of Volume 1:

$$\delta j_{goal}(\mathcal{M}) \preceq |a(\Pi_h u - u, u_{g,h}^*)| + |(f - \Pi_h f, u_{g,h}^*)| + |g||\pi_h u_h - u_h|.$$

For the second $\Pi_h u - u$ term, we first apply lemma 5.1 and then also introduce the continuous interpolation error. We get

$$\delta j_{goal}(\mathcal{M}) \preceq \int_\Omega \left([\frac{1}{\rho}\rho_S(H(u_{g,h}^*)) + |g|] \, |\pi_h u_h - u_h| + |u_{g,h}^*| \, |\pi_h f - f| \right) d\mathbf{x}.$$

$$[5.8]$$

It is then reasonable to try to minimize the RHS of this inequality instead of the LHS. But this still involves some difficulty due to the dependency of adjoint state $u^*_{g,h}$ with respect to $\mathcal{M}$. We shall further simplify our functional by freezing the adjoint state during a part of the algorithm. The idea is that when we change the parameter $\mathcal{M}$, the discrete adjoint $u^*_{g,h}$ is close to its (non-zero) continuous limit and is thus not much affected, in contrast to the interpolation errors $|\pi_h u_h - u_h|$ and $|\pi_h f - f|$. We then consider, for a given $\mathcal{M}_0$, the following optimum problem:

$$\min_{\mathcal{M}} \int_{\Omega} \left([\frac{1}{\rho}\rho_S(H(u^*_{g,\mathcal{M}_0})) + |g|] \, |\pi_h u_h - u_h| + |u^*_{g,\mathcal{M}_0}| \, |\pi_h f - f| \right) d\mathbf{x}.$$

This produces an optimum:

$$\mathcal{M}_{opt,\mathcal{M}_0} = Argmin_{\mathcal{M}} |tr(\mathcal{M}^{-1/2}$$
$$\times \left([\frac{1}{\rho}\rho_S H(u^*_{g,\mathcal{M}_0}) + |g|] |H_u| + |u^*_{g,\mathcal{M}_0}||H_f| \right) \mathcal{M}^{-1/2})|.$$

Observing that, in the integrand, the matrix

$$H_{goal,0} = [\frac{1}{\rho}\rho_S(H(u^*_{g,\mathcal{M}_0})) + |g|] \, |H_u| + |u^*_{g,\mathcal{M}_0}| \, |H_f|$$

is positive symmetric positive, we can apply the calculus of variation and get

$$\mathcal{M}_{opt,\mathcal{M}_0} = \mathcal{K}_1([\frac{1}{\rho}\rho_S(H(u^*_{g,\mathcal{M}_0})) + |g|] \, |H_u| + |u^*_{g,\mathcal{M}_0}| \, |H_f|),$$

where $\mathcal{K}_1$ is defined in [4.40] of corollary 4.2 of Volume 1. This solution can then be introduced in a fixed-point loop and will result in the solution

$$\mathcal{M}_{opt,goal} = \mathcal{K}_1([\frac{1}{\rho}\rho_S(H(u^*_{g,\mathcal{M}_{opt,goal}})) + |g|] \, |H_u| + |u^*_{g,\mathcal{M}_{opt,goal}}| \, |H_f|).$$

This expression is the analog of reference [4.16] derived in Chapter 4 for steady Euler flow. It permits after discretization to apply an anisotropic goal-oriented algorithm, for example, Algorithm 4.1 of this volume (the viscous version is given as Algorithm 5.2 in the sequel).

5.2.3. *Goal-oriented adaptation according to a second estimate*

We give now a development inspired by Michal et al. (2018a) and allowing the compensation of terms in the estimate, and therefore possibly giving a more accurate estimate than [5.8]. We define

$$\mathcal{F}^V(u) = \nabla u.$$

The analysis starts as follows:

$$|(g, u - u_h)| \approx \left| \int_\Omega \left[u^* \left(-\nabla \cdot (\mathcal{F}^V(u) - \mathcal{F}^V(\Pi_h u)) \right) \right. \right.$$
$$\left. \left. + \Pi_h f - f \right) + g \left(u - \Pi_h u \right) \right] \mathrm{d}\Omega \right|$$
$$\leq \left| \int_\Omega \sum_{i,j} \nabla_{x_i} \left[\frac{\partial \mathcal{F}^V}{\partial \nabla_{x_j} u} \nabla_{x_j} (u - \Pi_h u) \right] u^* \right.$$
$$\left. + u^* \left(\Pi_h f - f \right) + g \left(u - \Pi_h u \right) \mathrm{d}\Omega \right|.$$

Then, integrating by parts, we pass the derivatives on the adjoint state (again neglecting boundary terms):

$$|(g, u - u_h)| \leq \left| \int_\Omega \left[g + \sum_{i,j} \frac{\partial \mathcal{F}^V}{\partial \nabla_{x_j} u} \nabla_{x_i, x_j} u^* \right] (u - \Pi_h u) + u^* (\Pi_h f - f) \, \mathrm{d}\Omega \right|.$$

Then replacing $\mathcal{F}^V$, we get

$$|(g, u - u_h)| \leq \left| \int_\Omega \left[g + \sum_i \nabla_{x_i, x_i} u^* \right] (u - \Pi_h u) + u^* (\Pi_h f - f) \, \mathrm{d}\Omega \right|.$$

This error estimate is a weighted sum of interpolation errors in L^1-norm on the unknown and the RHS, where the weights depend on the Hessian of the adjoint state:

$$|(g, u - u_h| \leq \int_\Omega \left(\left| g + \sum_i \nabla_{x_i, x_j} u^* \right| |u - \Pi_h u| + |u^*| |\Pi_h f - f| \right) \mathrm{d}\Omega.$$

Similarly to the previous section, we consider, for a given $\mathcal{M}_0$, the following optimum problem:

$$\min_{\mathcal{M}} \int_{\Omega} \left(\left|g + \sum_i \nabla_{x_i,x_j} u^*\right| |u - \pi_h u| + |u^*| |\pi_h f - f| \right) d\Omega.$$

This produces an optimum (with frozen adjoint):

$$\mathcal{M}_{opt,\mathcal{M}_0} = Argmin_{\mathcal{M}} |tr(\mathcal{M}^{-1/2}$$
$$\times \left(\left|g + \sum_i \nabla_{x_i,x_j} u^*\right| |H_u| + |u^*_{g,\mathcal{M}_0}| |H_f|\right) \mathcal{M}^{-1/2})|.$$

(again an analog of reference [4.16] of section 4.4 of Chapter 4 of this volume) and completes the evaluation of the optimal metric. In this more globally linearized second estimate, we have more freedom for terms to compensate each other, thus obtaining a more accurate estimate.

We do not present in this chapter any numerical validation relying on the two above elliptic analyses, which we use as models for the Navier–Stokes applications. Chapter 6 of this volume presents such a validation for an extension of the first elliptic analysis.

5.3. Error analysis for Navier–Stokes problem

5.3.1. *Mesh adaptation problem statement*

We continue to assume that the purpose of the numerical problem is to evaluate the output functional j, but this time expressed through a steady Navier–Stokes system:

$$j = (g, W) \quad \text{where } W \text{ is the solution of problem [1.51] (Volume 1).} \quad [5.9]$$

More precisely, let us consider the continuous W and discrete W_h solutions of the continuous [1.51] and discrete [1.55] laminar Navier–Stokes models introduced in Chapter 1 of Volume 1. The problem addressed in this chapter is to *find the discrete mesh which minimizes the following functional error given a fixed number of vertices N:*

$$\delta j = (g, W - W_h),$$

where W is the solution of problem [1.51] and W_h is the solution of problem [1.55].

This section and the next ones are devoted to this error analysis and the optimal formulation of the mesh adaptation problem. Let ψ be a smooth test function of $\mathcal{V}$. Let W be the solution of problem [1.51] and W_h the solution of problem [1.55], and the continuous and discrete state equation is written as

$$(\Psi(W), \psi) = 0 \quad \text{and} \quad (\Psi_h(W_h), \Pi_h \psi) = 0,$$

where $\Pi_h \psi$ lies in $\mathcal{V}_h$ defined by relation [1.54]. We also introduce the continuous and the discrete adjoint states: W^* and W_h^*. The continuous adjoint system related to the objective functional is written as

$$W^* \in \mathcal{V}, \forall \psi \in \mathcal{V} : \left(\frac{\partial \Psi}{\partial W}(W)\psi, W^* \right) = (g, \psi). \qquad [5.10]$$

From functional analysis theory, a well-posed continuous adjoint system can be derived for any functional output as far as the linearized system is well posed. This, however, does not mean that any output functional leads to properly defined adjoint boundary conditions. Several works in the literature (Anderson and Venkatakrishnan 1999; Arian and Salas 1999; Bueno-Orovio et al. 2012; Castro et al. 2007) illustrate this problem and propose solutions, usually by adding auxiliary boundary terms to the Lagrangian functional. In Castro et al. (2007), it is concluded that for the compressible Navier–Stokes system, only functionals that involve the entire stress tensor at obstacle boundary are admissible. We assume here that [5.10] is well posed and gives a sufficiently smooth continuous adjoint state. The discrete adjoint system is written as

$$W_h^* \in \mathcal{V}_h, \forall \psi_h \in \mathcal{V}_h : \left(\frac{\partial \Psi_h}{\partial W}(W_h)\psi_h, W_h^* \right) = (g, \psi_h).$$

5.3.2. *Linearized error system*

In our error estimation problem presented in section 5.3.1, the approximation error can be decomposed into an implicit error term and an interpolation error term:

$$\delta j = (g, W - W_h) = (g, W - \Pi_h W) + (g, \Pi_h W - W_h).$$

The interpolation error can be easily estimated, corollary 4.1 of Volume 1, while the implicit error is solution of a discrete system that we derive in the following. Now, we assume that W_h can be made close enough to $\Pi_h W$ when $h \to 0$ in such a way that we can write

$$\left(\frac{\partial \Psi_h}{\partial W}(W_h)(W_h - \Pi_h W), \Pi_h \psi\right) \approx \left(\Psi_h(W_h) - \Psi_h(\Pi_h W), \Pi_h \psi\right)$$

as $h \to 0$.

Then, combining continuous and discrete systems, we can write similarly to relation [5.7] an equality linking implicit and interpolation errors, which is valid for all ψ:

$$\begin{aligned}
(\Psi_h(W_h) - \Psi_h(\Pi_h W), \Pi_h \psi) &= (\Psi_h(W_h), \Pi_h \psi) - (\Psi_h(\Pi_h W), \Pi_h \psi) \\
&= (\Psi(W), \Pi_h \psi) - (\Psi_h(\Pi_h W), \Pi_h \psi) \\
&= (\Psi(W) - \Psi_h(\Pi_h W), \Pi_h \psi)
\end{aligned}$$

We are then interested in the following error on the functional:

$$|(g, W_h - \Pi_h W)| \approx (\Psi_h(W_h) - \Psi_h(\Pi_h W), W_h^*) = (\Psi(W) - \Psi_h(\Pi_h W), W_h^*).$$

The right-hand side of the above relation is composed of the Euler term (see relation [1.55]), the Euler boundary term, the viscous term and the stabilization term. In the following, we neglect the boundary term (as in the previous analysis) and the stabilization term because of smoothness of functions W and W^* [3]. The method proposed here involves some heuristics. In particular, we assume that *the interpolate of the adjoint is close to the discrete adjoint*:

$$\Pi_h W^* \approx W_h^*.$$

Therefore

$$|(g, W_h - \Pi_h W)| \approx (\Psi(W) - \Psi_h(\Pi_h W), \Pi_h W^*).$$

5.3.3. *First estimate for Navier–Stokes problem*

The first a priori estimate that we propose is stated as follows:

---

[3] The boundary term analysis is given in Loseille et al. (2010a).

PROPOSITION 5.1.– Let us assume that $W \in \mathcal{V}$ and $\boldsymbol{\psi} \in \mathcal{V}$, $\boldsymbol{\psi} = (\psi_\rho, \psi_{\rho u_1}, \psi_{\rho u_2}, \psi_{\rho u_3}, \psi_{\rho E})^T$. Then, we have the following error bound for h sufficiently small:

$$|(\Psi(W) - \Psi_h(\Pi_h W), \Pi_h \boldsymbol{\psi})| \preceq \mathcal{E},$$

with:

$$\mathcal{E} = \int_\Omega |\nabla \boldsymbol{\psi}| \cdot |\mathcal{F}^E(W) - \Pi_h \mathcal{F}^E(W)| \, d\Omega$$

$$+ \left(d + \frac{1}{3}\right) K_d \sum_{i=1}^{d} \int_\Omega \mu \, |\rho_H(\psi_{\rho u_i})| \, |u_i - \Pi_h u_i| \, d\Omega$$

$$+ \frac{1}{3} K_d \sum_{i=1}^{d} \sum_{\substack{j=1 \\ j \neq i}}^{d} \int_\Omega \mu \, |\rho_H(\psi_{\rho u_i})| \, |u_j - \Pi_h u_j| \, d\Omega$$

$$+ K_d \int_\Omega \lambda \, |\rho_H(\psi_{\rho E})| \, |T - \Pi_h T| \, d\Omega$$

$$+ (d + \frac{1}{3}) K_d \sum_{i=1}^{d} \int_\Omega \mu \, |\rho_H(\psi_{\rho E})| \, |u_i| \, |u_i - \Pi_h u_i| \, d\Omega$$

$$+ \frac{1}{3} K_d \sum_{i=1}^{d} \sum_{\substack{j=1 \\ j \neq i}}^{d} \int_\Omega \mu \, |\rho_H(\psi_{\rho E})| \, |u_i| \, |u_j - \Pi_h u_j| \, d\Omega$$

$$+ \frac{5}{3} \sum_{i=1}^{d} \sum_{\substack{j=1 \\ j \neq i}}^{d} \int_\Omega \mu \left| \frac{\partial (\Pi_h \psi_{\rho E})}{\partial x_j} \left((u_j - \Pi_h u_j) \frac{\partial u_i}{\partial x_i} - (u_i - \Pi_h u_i) \frac{\partial u_j}{\partial x_i} \right) \right| d\Omega,$$

and $K_d = 3$ in two dimensions, $K_d = 6$ in three dimensions, and $A \preceq B$ holds for a majoration asymptotically valid, that is, $A \leq B + o(A)$ when mesh size tends to zero. Expression $|\rho_H(\varphi)|$ holds for spectral radius of the Hessian of φ.

The proof of this proposition is very technical and is given in Belme et al. (2019).
□

This main result can be written in a more convenient way to facilitate its implementation by gathering interpolation error terms on the primitive variables. We give now the 2D and the 3D re-writing of it.

COROLLARY 5.1.– Let us assume that $W \in \mathcal{V}$ and $\psi \in \mathcal{V}$, $\psi = (\psi_\rho, \psi_{\rho u_1}, \psi_{\rho u_2}, \psi_{\rho E})^T$. Then, in two dimensions ($K_d = 3$), we have the following error bound for h sufficiently small:

$$|(\Psi(W) - \Psi_h(\Pi_h W), \Pi_h \psi)| \preceq \int_\Omega G_{\mathcal{F}^E} \left|\mathcal{F}^E(W) - \Pi_h \mathcal{F}^E(W)\right| d\Omega$$

$$+ \int_\Omega G_{u_1} |u_1 - \Pi_h u_1| d\Omega + \int_\Omega G_{u_2} |u_2 - \Pi_h u_2| d\Omega$$

$$+ \int_\Omega G_T |T - \Pi_h T| d\Omega,$$

with the coefficients

$$G_{\mathcal{F}^E} = |\nabla \psi|$$

$$G_{u_1} = \mu \left(7 |\rho_H(\psi_{\rho u_1})| + |\rho_H(\psi_{\rho u_2})| \right.$$

$$\left. + \left(7|u_1| + |u_2| \right) |\rho_H(\psi_{\rho E})| + \frac{5}{3} |\omega_{u_2,z}| \right)$$

$$G_{u_2} = \mu \left(|\rho_H(\psi_{\rho u_1})| + 7 |\rho_H(\psi_{\rho u_2})| \right.$$

$$\left. + \left(|u_1| + 7|u_2| \right) |\rho_H(\psi_{\rho E})| + \frac{5}{3} |\omega_{u_1,z}| \right)$$

$$G_T = 3\lambda |\rho_H(\psi_{\rho E})|$$

and the ω vector defined by

$$\nabla u_i \times \nabla \psi_{\rho E} = \omega_{u_i} = \left(\omega_{u_i,x}, \omega_{u_i,y}, \omega_{u_i,z} \right)^T,$$

where only the last component is not zero. □

COROLLARY 5.2.– Let us assume that $W \in \mathcal{V}$ and $\psi \in \mathcal{V}$, $\psi = (\psi_\rho, \psi_{\rho u_1}, \psi_{\rho u_2}, \psi_{\rho u_3}, \psi_{\rho E})^T$. Then, in three dimensions ($K_d = 6$), we have the following error bound for h sufficiently small:

$$|(\Psi(W) - \Psi_h(\Pi_h W), \Pi_h \psi)| \preceq \int_\Omega G_{\mathcal{F}^E} \left|\mathcal{F}^E(W) - \Pi_h \mathcal{F}^E(W)\right| d\Omega$$

$$+ \int_\Omega G_{u_1} |u_1 - \Pi_h u_1| d\Omega + \int_\Omega G_{u_2} |u_2 - \Pi_h u_2| d\Omega$$

$$+ \int_\Omega G_{u_3} |u_3 - \Pi_h u_3| d\Omega + \int_\Omega G_T |T - \Pi_h T| d\Omega,$$

with the coefficients

$$G_{\mathcal{F}E} = |\nabla \psi|$$
$$G_{u_1} = \mu \left(20\,|\rho_H(\psi_{\rho u_1})| + 2\,|\rho_H(\psi_{\rho u_2})| + 2\,|\rho_H(\psi_{\rho u_3})|\right)$$
$$+ \mu \left(\left(20|u_1| + 2|u_2| + 2|u_3|\right) |\rho_H(\psi_{\rho E})| + \tfrac{5}{3}|\omega_{u_3,y} - \omega_{u_2,z}|\right)$$
$$G_{u_2} = \mu \left(2\,|\rho_H(\psi_{\rho u_1})| + 20\,|\rho_H(\psi_{\rho u_2})| + 2\,|\rho_H(\psi_{\rho u_3})|\right)$$
$$+ \mu \left(\left(2|u_1| + 20|u_2| + 2|u_3|\right) |\rho_H(\psi_{\rho E})| + \tfrac{5}{3}|\omega_{u_1,z} - \omega_{u_3,x}|\right)$$
$$G_{u_3} = \mu \left(2\,|\rho_H(\psi_{\rho u_1})| + 2\,|\rho_H(\psi_{\rho u_2})| + 20\,|\rho_H(\psi_{\rho u_3})|\right)$$
$$+ \mu \left(\left(2|u_1| + 2|u_2| + 20|u_3|\right) |\rho_H(\psi_{\rho E})| + \tfrac{5}{3}|\omega_{u_2,x} - \omega_{u_1,y}|\right)$$
$$G_T = 6\,\lambda\,|\rho_H(\psi_{\rho E})|$$

and the ω vector defined by

$$\nabla u_i \times \nabla \psi_{\rho E} = \boldsymbol{\omega}_{u_i} = (\omega_{u_i,x}, \omega_{u_i,y}, \omega_{u_i,z})^T. \qquad \square$$

For using the estimate in a mesh adaptation loop, we seek for the optimal mesh that minimizes the above error model. It is useful to note that the error model can be written under the compact form:

$$|(\Psi(W) - \Psi_h(\Pi_h W), \Pi_h \psi)| \preceq \int_\Omega \sum_k G_k(\mu, \lambda, U, \nabla \psi, |\rho_H(\psi)|) \left|S_k(W)\right.$$
$$\left. - \Pi_h S_k(W)\right| \,\mathrm{d}\Omega.$$

In other words, the error model is a sum of interpolation errors weighted by algebraic functions.

We can reformulate the discrete mesh adaptation problem introduced in section 5.3.1 in the continuous mesh framework: *find the continuous mesh* $\mathbf{M}_{opt}$ *which minimizes the following functional error given a fixed complexity* $\mathcal{N}$:

$$\delta j = (g, W - W_h) \approx (g, W - \Pi_h W) + (\Psi(W) - \Psi_h(\Pi_h W), \Pi_h W^*)$$
$$\preceq \int_\Omega \sum_k |g_k|\,|W_k - \pi_\mathcal{M} W_k|\,\mathrm{d}\Omega$$
$$+ \int_\Omega \sum_k G_k(\mu, \lambda, U, \nabla W^*, |\rho_H(W^*)|) \left|S_k(W)\right.$$
$$\left. - \pi_\mathcal{M} S_k(W)\right| \,\mathrm{d}\Omega,$$

where W is the solution of problem [1.51], W_h is the solution of problem [1.55] and W^* is the solution of problem [5.10].

The above relation is a weighted sum of interpolation errors. By introducing the following positive symmetric matrix

$$\mathbf{H}_{go1}(\mathbf{x}) = \sum_k |g_k| |H_{W_k}| + \sum_k G_k(\mu, \lambda, U, \nabla W^*, |\rho_H(W^*)|) |H_{S_k(W)}|$$

where $|H_{W_k}|$ and $|H_{S_k(W)}|$ are the absolute value of the Hessian of fields W_k and $S_k(W)$, and using the definition of the continuous interpolation error given by corollary 4.1 of Volume 1, we can state the following error estimate on continuous mesh $\mathbf{M} = (\mathcal{M}(\mathbf{x}))_{\mathbf{x}\in\Omega}$:

$$\delta j \approx \mathbf{E}_{go1}(\mathbf{M}) = \int_\Omega \text{trace}\left(\mathcal{M}^{-\frac{1}{2}}(\mathbf{x})\,\mathbf{H}_{go1}(\mathbf{x})\,\mathcal{M}^{-\frac{1}{2}}(\mathbf{x})\right)\,\mathrm{d}\Omega.$$

5.3.4. *Second estimate for Navier–Stokes problem*

We adapt the second estimate for elliptic case (see section 5.2.3). In Michal et al. (2017, 2018a,b), this type of analysis was proposed for Navier–Stokes and has proved to be more efficient on the ONERA M6 case, among other cases, than the first estimate of previous section. While in Michal et al. (2017, 2018a,b) an approximation was made by using only the Hessian of the Mach number variable, we propose here to introduce ingredients of the first estimate in order to derive a sharper viscous goal-oriented error estimate. Moreover, this analysis can be done on the conservative variable to end-up only with five terms in the error estimation. Both inviscid and viscous fluxes are linearized with respect to the solution W and its gradients ∇W and integrating by parts (and omitting the boundary terms), we obtain

$$|(g, W_h - \Pi_h W)| \approx \left| \int_\Omega W^* \nabla \cdot (\mathcal{F}^E(W) - \mathcal{F}^E(\Pi_h W)) - \nabla \cdot (\mathcal{F}^V(W) \right.$$
$$\left. - \mathcal{F}^V(\Pi_h W))\,\mathrm{d}\Omega \right|$$
$$\leq \left| \int_\Omega \left(\sum_i \nabla_{x_i} \left[\frac{\partial \mathcal{F}_i^E}{\partial W}(W - \Pi_h W) \right] \right. \right.$$

$$+ \sum_{i,j} \nabla_{x_i} \left[\frac{\partial \mathcal{F}_i^V}{\partial \nabla_{x_j} W} \nabla_{x_j} (W - \Pi_h W) \right] \Big) W^* \mathrm{d}\Omega \Big|$$

$$\leq \int_\Omega \Big| \sum_i \frac{\partial \mathcal{F}_i^E}{\partial W} \nabla_{x_i} W^* + \sum_{i,j} \frac{\partial \mathcal{F}_i^V}{\partial \nabla_{x_j} W} \nabla_{x_i, x_j} W^* \Big| \big| W - \Pi_h W \big| \mathrm{d}\Omega. \qquad [5.11]$$

This implicit error estimate is a weighted sum of interpolation errors in L^1-norm on the conservative variables where the weights depend on the gradient and the Hessian of the adjoint state and on the convective and viscous fluxes. We just have to add the interpolation error term to have the approximation error estimate:

$$|J(W) - J(W_h)| \leq \int_\Omega \Big| \frac{\partial j}{\partial W} + \sum_i \frac{\partial \mathcal{F}_i^E}{\partial W} \nabla_{x_i} W^*$$

$$+ \sum_{i,j} \frac{\partial \mathcal{F}_i^V}{\partial \nabla_{x_j} W} \nabla_{x_i, x_j} W^* \Big| \big| W - \Pi_h W \big| \mathrm{d}\Omega.$$

For practical implementation, we give the expression of the weights in 3D for each conservative variable:

$$|J(W) - J(W_h)| \leq \int_\Omega \Big(\sum_{W_k = \rho, \rho u_1, \rho u_2, \rho u_3, \rho E} |G_{W_k}| |W_k - \Pi_h W_k| \Big) \mathrm{d}\Omega,$$

with

$$G_\rho = \frac{\partial j}{\partial \rho} + \sum_i \frac{\partial \mathcal{F}_i^E}{\partial \rho} \cdot \nabla_{x_i} W^* - u_1 f_{\rho u_1}(W, H_{W^*})$$
$$- u_2 f_{\rho u_2}(W, H_{W^*}) - u_3 f_{\rho u_3}(W, H_{W^*})$$
$$+ (u^2 + v^2 + w^2 - E) f_{\rho E}(W, H_{W^*})$$

$$G_{\rho u_1} = \frac{\partial j}{\partial \rho u_1} + \sum_i \frac{\partial \mathcal{F}_i^E}{\partial \rho u_1} \cdot \nabla_{x_i} W^* + f_{\rho u_1}(W, H_{W^*}) - u f_{\rho E}(W, H_{W^*})$$

$$G_{\rho u_2} = \frac{\partial j}{\partial \rho u_2} + \sum_i \frac{\partial \mathcal{F}_i^E}{\partial \rho u_2} \cdot \nabla_{x_i} W^* + f_{\rho u_2}(W, H_{W^*}) - v f_{\rho E}(W, H_{W^*})$$

$$G_{\rho u_3} = \frac{\partial j}{\partial \rho u_3} + \sum_i \frac{\partial \mathcal{F}_i^E}{\partial \rho u_3} \cdot \nabla_{x_i} W^* + f_{\rho u_3}(W, H_{W^*}) - w\, f_{\rho E}(W, H_{W^*})$$

$$G_{\rho E} = \frac{\partial j}{\partial \rho E} + \sum_i \frac{\partial \mathcal{F}_i^E}{\partial \rho E} \cdot \nabla_{x_i} W^* + f_{\rho E}(W, H_{W^*}),$$

where we have

$$f_{\rho u_1}(W, H_{W^*}) = \frac{1}{3}\frac{\mu + \mu_t}{\rho}\bigg(4\,(\rho u_1^*)_{xx} + 3\,(\rho u_1^*)_{yy} + 3\,(\rho u_1^*)_{zz}$$
$$+ (\rho u_2^*)_{xy} + (\rho u_3^*)_{xz} \ + \ 4\,u_1\,(\rho E^*)_{xx} + 3\,u_1\,(\rho E^*)_{yy}$$
$$+ 3\,u_1\,(\rho E^*)_{zz} + u_2\,(\rho E^*)_{xy} + u_3\,(\rho E^*)_{xz}\bigg)$$

$$f_{\rho u_2}(W, H_{W^*}) = \frac{1}{3}\frac{\mu + \mu_t}{\rho}\bigg((\rho u_1^*)_{xy} + 3\,(\rho u_2^*)_{xx} + 4\,(\rho u_2^*)_{yy} + 3\,(\rho u_2^*)_{zz}$$
$$+ (\rho u_3^*)_{yz} + \ u_1\,(\rho E^*)_{xy} + 3\,u_2\,(\rho E^*)_{xx} + 4\,u_2\,(\rho E^*)_{yy}$$
$$+ 3\,u_2\,(\rho E^*)_{zz} + u_3\,(\rho E^*)_{yz}\bigg)$$

$$f_{\rho u_3}(W, H_{W^*}) = \frac{1}{3}\frac{\mu + \mu_t}{\rho}\bigg((\rho u_1^*)_{xz} + (\rho u_2^*)_{yz} + 3\,(\rho u_3^*)_{xx} + 3\,(\rho u_3^*)_{yy}$$
$$+ 4\,(\rho u_3^*)_{zz} + u_1\,(\rho E^*)_{xz} + u_2\,(\rho E^*)_{yz} + 3\,u_3\,(\rho E^*)_{xx}$$
$$+ 3\,u_3\,(\rho E^*)_{yy} + 4\,u_3\,(\rho E^*)_{zz}\bigg)$$

$$f_{\rho E}(W, H_{W^*}) = \frac{\lambda + \lambda_t}{\rho}\Big((\rho E^*)_{xx} + (\rho E^*)_{yy} + (\rho E^*)_{zz} \Big).$$

Again, we can re-use the interpolation error analysis, this time with

$$\mathbf{H}_{go2}(\mathbf{x}) = \sum_{W_k = \rho, \rho u_1, \rho u_2, \rho u_3, \rho E} |G_{W_k}||H_{W_k}|, \qquad [5.12]$$

to obtain a second estimate:

$$\delta j \approx \mathbf{E}_{go2}(\mathbf{M}) = c_d \int_\Omega \text{trace}\left(\mathcal{M}^{-\frac{1}{2}}(\mathbf{x})\, \mathbf{H}_{go2}(\mathbf{x})\, \mathcal{M}^{-\frac{1}{2}}(\mathbf{x}) \right) \mathrm{d}\Omega.$$

For RANS computations, this estimate should also take into account in the estimate the turbulent closure equations. Our option here is to neglect them. The influence of turbulence modeling is then limited to accounting the variables μ_t and λ_t in the mean flow equations.

5.3.5. *Optimal goal-oriented continuous mesh*

Equipped with the estimate

$$\delta j \approx \mathbf{E}_{go}(\mathbf{M}) \;=\; \mathbf{E}_{go1}(\mathbf{M}) \;\; \text{or} \;\; \mathbf{E}_{go2}(\mathbf{M}),$$

it is possible to set the well-posed global optimization problem of finding the optimal continuous mesh $\mathbf{M}_{go}$ minimizing continuous interpolation error $\mathbf{E}_{go}(\mathcal{M})$:

$$\text{Find } \mathbf{M}_{go} = \min_{\mathbf{M}} \mathbf{E}_{go}(\mathbf{M}) \quad \text{under the constraint} \quad \mathcal{C}(\mathbf{M}) = N. \qquad [5.13]$$

We seek for the optimal continuous mesh $\mathbf{M}_{go}$ solution of problem [5.13]. In a similar way to Chapter 4 of this volume, solving the optimality conditions provides the *optimal goal-oriented continuous mesh* $\mathbf{M}_{go} = (\mathcal{M}_{go}(\mathbf{x}))_{\mathbf{x} \in \Omega}$ defined point-wise as follows (we recall that d is the space dimension):

$$\mathcal{M}_{go}(\mathbf{x}) = N^{\frac{2}{d}} \left(\int_{\Omega} (\det \mathbf{H}_{go}(\bar{\mathbf{x}}))^{\frac{1}{d+2}} d\bar{\mathbf{x}} \right)^{-\frac{2}{d}} \left(\det \mathbf{H}_{go}(\mathbf{x}) \right)^{-\frac{1}{d+2}} \mathbf{H}_{go}(\mathbf{x}). \qquad [5.14]$$

The corresponding optimal error is written as

$$\mathbf{E}_{go}(\mathbf{M}_{go}) = = 3\, N^{-\frac{2}{d}} \left(\int_{\Omega} (\det \mathbf{H}_{go}(\mathbf{x}))^{\frac{1}{d+2}} d\mathbf{x} \right)^{\frac{2+d}{d}}, \qquad [5.15]$$

where the exponent of $\mathcal{N}$ illustrates the second-order accuracy of the method.

5.4. From theory to practice

The continuous mesh adaptation problem takes the form of the following *continuous optimality system*:

$$W \in \mathcal{V}, \; \forall \psi \in \mathcal{V}, \qquad (\Psi(\mathbf{M}, W), \psi) = 0 \qquad \text{"Navier–Stokes system"}$$

$$W^* \in \mathcal{V}, \; \forall \psi \in \mathcal{V}, \; \left(\frac{\partial \Psi}{\partial W}(\mathbf{M}, W)\psi, W^* \right) = (g, \psi) \quad \text{"Adjoint system"}$$

$$\mathbf{M} = \mathbf{M}_{go} \qquad \text{"Adapted continuous mesh"}.$$

In practice, it is necessary to approximate the above three-field coupled system by the *discrete optimality system*:

$$W \in \mathcal{V}_h, \, \forall \psi_h \in \mathcal{V}_h, (\Psi_h(\mathcal{H}, W_h), \psi_h) = 0$$

"Discrete Navier–Stokes system"

$$W_h^* \in \mathcal{V}_h, \, \forall \psi_h \in \mathcal{V}_h, \left(\frac{\partial \Psi}{\partial W}(\mathcal{H}, W_h)\psi_h, W_h^*\right) = (g, \psi_h)$$

"Discrete Adjoint system"

$\mathcal{H} = \mathcal{H}_{go} = \mathcal{H}(\mathbf{M}_{go})$ "Discrete adapted mesh".

The discrete optimality system is solved using the fixed-point mesh adaptation Algorithm 5.1. *In practice, Algorithm 5.1 should be called by Algorithm 6.1 of Volume 1 for obtaining mesh-converged results.*

Algorithm 5.1. Viscous goal-oriented mesh adaptation loop for steady flows

---

Initial mesh and solution $(\mathcal{H}_{go}^1, \mathcal{S}_0^1)$ and set targeted functional j_h and complexity N

 # Adaptive loop to converge the steady-state solution and its associated optimal adapted discrete mesh
For $i = 1, n_{adap}$

1) W_h^i = Compute discrete state solution of the discrete Navier–Stokes system from pair $(W_{h,0}^i, \mathcal{H}_{go}^i)$;

2) $W_h^{*,i}$ = Compute discrete adjoint state solution of the discrete adjoint system from $(j_h, W_h^i, \mathcal{H}_{go}^i)$;

3) $\mathbf{M}_{go}^i$ = Compute the optimal goal-oriented metric field from $(N, W_h^{*,i}, W_h^i, \mathcal{H}_{go}^i)$;

4) $\mathcal{H}_{go}^{i+1}$ = Generate the new adapted mesh which is unit w.r.t. $\mathbf{M}_{go}^i$ from $(\mathbf{M}_{go}^i, \mathcal{H}_{go}^i)$;

5) W_0^{i+1} = Interpolate the previous discrete state solution on the new mesh from $(\mathcal{H}_{go}^i, W^i, \mathcal{H}_{go}^{i+1})$;

EndFor

The discrete state equation is modeled with `Wolf` code as in section 1.2.11 of Volume 1. With regard to the adjoint state computation, the matrix of the linear system is simply the transpose of the implicit state equation matrix without the time derivative term. The right hand-side of the system is the exact differentiation with respect to state of the chosen functional (for instance, drag, lift and so on):

$$\frac{\partial \Psi_h}{\partial W_h}^T W_h^* = \frac{\partial j}{\partial W_h}.$$

In particular, for viscous flows, μ and the stress tensor τ are exactly differentiated. To solve the adjoint system, we use a restarted GMRES preconditioned with SGS relaxation. Note that it is important to solve the adjoint linear system to machine precision.

5.4.1. *Computation of the optimal continuous mesh*

The expression of the optimal goal-oriented metric field is either given by relations [5.11]–[5.14] where algebraic functions G_k and S_k are given in corollaries 5.1 and 5.2 where ψ is replaced by the adjoint W^*, or by using section 5.3.4. This expression involves the state and the adjoint state, and their derivatives (first and second). In practice, these terms are approximated by the discrete states and a derivative recovery is applied to get gradients and Hessians. The recovery method is based on the double L^2-projection formula described in Chapter 5 of Volume 1. We then obtain a discrete metric field defined at vertices of the mesh. The discrete adjoint state W_h^* is taken to represent the adjoint state W^*. The gradient of the adjoint state ∇W^* is replaced by $\nabla_R W_h^*$, where ∇_R stands for the operator that recovers numerically the first derivatives of an initial piecewise linear solution field. The Hessian is obtained by applying the recovery operator two times: $H_R = \nabla_R \circ \nabla_R$. Then, $|\rho_H(W^*)|$ is obtained from $|\rho_H(W_h^*)|$, which is evaluated by computing the maximal (in absolute value) eigenvalue of $H_R(W_h^*)$. Similarly, the discrete state W_h is taken to represent the state W and to compute the Euler fluxes, velocity components and temperature (i.e. each term $S_k(W)$ is replaced by $S_k(W_h)$). The Hessians of the $S_k(W_h)$ are obtained using the recovery operator H_R and the absolute values of the Hessians are computed by taking the absolute values of the eigenvalues.

5.5. An example of application to a turbulent flow

The goal-oriented mesh adaptation has been applied in Alauzet and Frazza (2020) to the high-lift version of the NASA CRM (HL-CRM) geometry used for the 3rd AIAA CFD High Lift Prediction Workshop (Rumsey et al. 2018) (HLPW3). A preliminary series of computations of this case showed that the first laminar estimate did not give good adaptation. We have retained in this chapter what appeared to be the two best error estimates:

– the *feature-based error estimate* controlling the interpolation of the local Mach number in L^4-norm;

– the *goal-oriented error estimate* given by relations [5.11]–[5.14].

For each error estimate:

– at most $n_{adap} = 20$ mesh adaptation iterations are performed at each fixed complexity and for the convergence study we consider five complexities (320,000; 640,000; 1,280,000; 2,560,000; 5,120,000);

– the convergence study is started with an initial mesh composed of 229,263 vertices, It is a very coarse inviscid mesh without any boundary layer or any specific refinement for viscous flows, in other word no meshing guidelines, thus very easy and quick to generate. Starting from this coarse and clearly unresolved mesh aims at illustrating the non-dependency of the mesh-adaptive solution platform to the initial mesh;

– for the given complexities, the final adapted meshes involve between 0.7M vertices and 10.2M vertices.

In Alauzet and Frazza (2020), the results obtained with the mesh-adaptive solution platform are compared to all the results obtained during the workshop and this allows to observe a similar quality of prediction between the workshop fine mesh results (205M vertices) and the mesh adaptive calculation with 10.2M vertices adapted mesh. The feature-based option is clearly less efficient than the goal-oriented option in this case but it produces similar results to the best practice meshes of workshop, with seven times less vertices.

For mesh convergence, we have combined Algorithm 5.1 with Algorithm 6.1 of Volume 1. The convergence of the lift value for the viscous goal-oriented error estimate throughout the whole mesh-convergence analysis is shown in Figure 5.1 (left). It shows the evolution of the lift for each computation in red (i.e. each adaptation at each complexity) and the final retained value obtained for each complexity in blue. We note that a lot is done on coarse adapted meshes (which is cheap), while the minimum number of iterations is done on the finer adapted meshes. Again, converging on coarse meshes is advantageous and enables early capturing of the solution. Figure 5.1 (right) compares the convergence of the lift for the solution mesh-adaptive platform (red and blue lines) with respect to all workshop entries (green lines). This again emphasizes the early capturing of the lift value on coarse adapted meshes in comparison to the meshes used for the workshop.

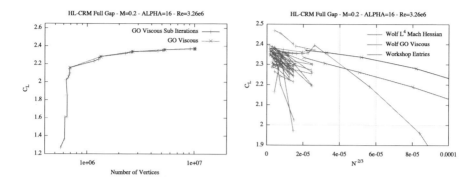

Figure 5.1. *HL-CRM $16°$ case. Left: Convergence history of the total lift value C_L for the viscous goal-oriented error estimate throughout the whole mesh-convergence analysis. In red, the convergence of the total lift at each complexity and, in blue, the global convergence of the total lift by retaining the final lift value for each complexity. Right: Convergence of the total lift for the solution mesh-adaptive platform with the feature-based error estimate (red line) and the viscous goal-oriented error estimate (blue line) with respect to all workshop entries (green lines)*

Now, we illustrate in a series of figures the favorable behavior of adaptation for capturing some important non-trivial details in the flow. The 5M vertices adapted mesh obtained with the viscous goal-oriented error estimate is used. We also show the local Mach number of the solution for each of these cuts to display all the high-lift flow features that are automatically capture by the mesh-adaptive solution process. Figure 5.2 displays a global view of the HL-CRM with a cut plane at $x = 50$ to emphasize the wake refinement. The medium and fine meshes of the workshop are refined isotropically in a large rectangular region hoping that the wake will be there for all angles of attack. The adapted mesh focuses only in the current wake and refine that region appropriately and anisotropically. For instance, we can clearly see the refinement of the four tip vortices coming from the main wing and the two flaps. Figures 5.3 and 5.4 exhibit close-up views of the slat and the flap for the cut plane $y = 15.5$. We observe that the adapted mesh focuses in the boundary layer which is very thin, the wakes of the slat and main wing that create shear layers, the wake – boundary layer merging and the flow separation over the flap, and also the separated cove flows. It is evident that meshing guidelines ask for a large number of layers in the boundary layer mesh to have some refinement far from the body in order to capture some of the shear layer. But, a shear layer located further above the body will render inefficient that strategy. Figures 5.5 and 5.6 show two details of the geometry that may impact the overall flow: the tip vortex emitted by the slat and the gap between the two flaps. First, in Figure 5.5, we observe that the adapted mesh refines a lot more the surface mesh and this mesh is highly anisotropic. Moreover, ridges of the geometry are a lot more refined because they can be at the intersection of two boundary layers or sources of detached vortices. We can also see in the adapted mesh and in the solution the wing tip vortex of the slat that interacts with the boundary layer of the main wing and detaches the flow there. Similarly, in Figure 5.6, the vortex that runs over the gap and that completely detaches the flow in that region and its interactions with the shear layers is accurately captured.

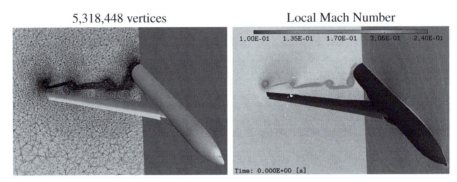

Figure 5.2. *HL-CRM $16°$ case. Cut plane $x = 50$. The 5M vertices adapted mesh obtained with the viscous goal-oriented error estimate (left) and the associated local Mach number solution (right). For a color version of this figure, see www.iste.co.uk/dervieux/ meshadaptation2*

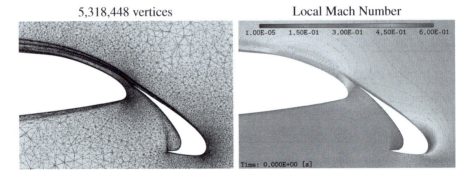

Figure 5.3. *HL-CRM $16°$ case. Cut plane $y = 15.5$ (near the flap). The 5M vertices adapted mesh obtained with the viscous goal-oriented error estimate (left) and the associated local Mach number solution (right). For a color version of this figure, see www.iste.co.uk/dervieux/meshadaptation2*

In conclusion, the more complex the geometry the more efficient should be the adaptive process as it able to automatically and accurately capture all the physical features associated with these geometry details, details which are very hard to mesh accurately with classical mesh generation methods.

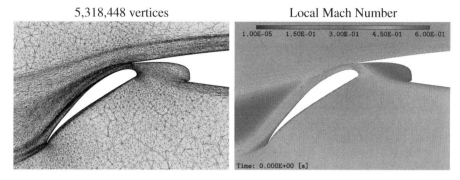

Figure 5.4. *HL-CRM* $16°$ *case. Cut plane* $y = 15.5$ *(near the slat). The 5M vertices adapted mesh obtained with the viscous goal-oriented error estimate (left) and the associated local Mach number solution (right). For a color version of this figure, see www.iste.co.uk/dervieux/meshadaptation2*

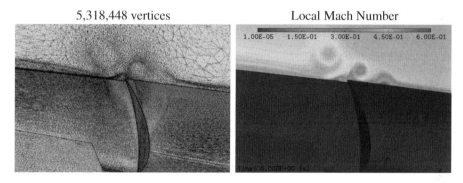

Figure 5.5. *HL-CRM* $16°$ *case. Cut plane in the region where the slat tip vortex interacts with the main wing. The 5M vertices adapted mesh obtained with the viscous goal-oriented error estimate (left) and the associated local Mach number solution (right). For a color version of this figure, see www.iste.co.uk/dervieux/meshadaptation2*

5.6. Conclusion

This chapter focuses on the extension to Navier–Stokes of the goal-oriented method developped for Euler flow in the previous chapter. We have developed a cautious error analysis from the Euler one. But this analysis neglects the event of term compensation. Beside this, we have described a second goal-oriented analysis applying directly the linearization of the Navier–Stokes equations. These two goal-oriented options are supposed to produce a better convergence than the third, less sophisticated, L^p feature-based adaptation.

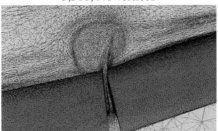 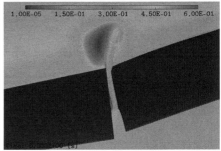

Figure 5.6. *HL-CRM* $16°$ *case. Cut plane* $x = 38$ *(near the gap between the flaps). The 5M vertices adapted mesh obtained with the viscous goal-oriented error estimate (left) and the associated local Mach number solution (right). For a color version of this figure, see www.iste.co.uk/dervieux/meshadaptation2*

The three corresponding fixed point algorithms have been inserted in the mesh-convergent loop. This loop is essential for the evaluation of the efficiency of a mesh adaptation method. It appeared that the first – analysis-based – goal-oriented approach, although producing second-order convergence for Euler flows, was in our experiments unable to second-order converge with median size meshes.

In the proposed numerical illustration, the L^4 feature-based and the linearized goal-oriented error estimates are compared on a high-lift workshop case. The viscous goal-oriented error estimate is clearly superior to the feature-based error estimate. Thus, it is a better choice if the adjoint state is available. With regard to the accuracy of the obtained results with the mesh adaptation process, the benefits are clear using the goal-oriented error estimate. The lift prediction agrees with the ones obtained during the workshop on the x-fine meshes composed of 206M vertices, but here the mesh size is between 5M and 10M vertices to reach that accuracy, that is, the size of the workshop coarse mesh! For the feature-based error estimate, the prediction is a bit lower and corresponds to the prediction on the fine workshop mesh (70M vertices). It will need one more level of complexity to see if it reaches the x-fine mesh prediction (note that in 2D all error estimates converge toward the same answer). In conclusion, anisotropic mesh adaptation is able to provide accurate high-lift prediction on meshes with a size similar to the coarse mesh used during the workshop. It also captures accurately all flow features and importantly all flow features that are created by geometric details. Thus, the more complex the geometry, the more efficient the adaptative process should be.

5.7. Notes

A preliminary formulation of the elliptic analysis was given in Belme (2011) and more complete derivation and extension to laminar Navier–Stokes is presented in Belme et al. (2019).

The 3D computations are presented in more details in Alauzet and Frazza (2020).

Other works. Anisotropic error estimates for elliptic models have been studied by several authors and, in particular, in Formaggia and Perotto (2001, 2003); Formaggia et al. (2004).

Acknowledgments. The work reported here was partially supported by funding from The Boeing Company with technical monitor Todd Michal. It was also partially supported by funding from Safran Tech with technical monitors Frédéric Feyel and Vincent Brunet.

6

Norm-Oriented Formulations

The previous chapters describe two important approaches for anisotropic mesh adaptation for PDEs, namely the feature-based approach and the goal-oriented approach. This chapter presents a third approach, the norm-oriented approach. The norm-oriented formulation extends the goal-oriented formulation since it is equation-based and uses an adjoint. At the same time, the norm-oriented formulation somewhat supersedes the goal-oriented approach since it is basically a *solution-convergent* method. Indeed, goal-oriented methods solely rely on the reduction of the error in evaluating a chosen scalar output with the consequence that, as mesh size is increased (more degrees of freedom), only this output is proven to tend to its continuous analog while the solution field itself may not converge. A remarkable quality of goal-oriented metric-based adaptation is the mathematical formulation of the mesh adaptation problem under the form of the optimization, in the well-identified set of metrics, of a well-defined functional. With the norm-oriented approach, this latter advantage is amplified by searching in the same well-identified set of metrics, the *minimum of a norm of the approximation error*. This norm then converges to zero when mesh size is increased. The type of norm is prescribed by the user and the method also addresses the case of multi-objective adaptation like, for example, in aerodynamics, adaptating the mesh for drag, lift and moment in one shot. Numerical examples for the Poisson problem and compressible flows are used for demonstrating the method.

6.1. Introduction

Goal-oriented methods have strongly impacted mesh adaptation applications. However, due to its formulation, a goal-oriented method has two inherent limitations. First, it does not naturally extend to several scalar outputs. This "multi-target" issue (explained in the sequel) is well known and a proposition for addressing it is made in

Hartman (2008). Second, since the goal-oriented method is specialized to a given scalar output, the features of the solution field that are not influencing this output might be neglected by the automatic mesh improvement. A goal-oriented method provides the convergence of the approximate chosen output to its continuous analog. But generally, convergence does not hold for the whole flow field itself. To clarify this point, let us consider the mesh adaptative computation of a sonic boom footprint at the ground as presented in section 4.6.1 of this volume. The functional depends only on the pressure at the ground level. Now, many details of the flow on the upper part of the aircraft do not influence the pressure at the ground level. This vanishing influence is taken into account by the adjoint state, which also vanishes on these upper regions. Then, in these regions, the adapted mesh is not refined and the approximation of the flow field does not converge when the prescribed mesh size is increased. Another limitation of a goal-oriented method is related to the scalar character of the output to be best approximated. It leads to the use of integrals of solution fields for example $(u - u_h, g)$, u being the exact solution, u_h its approximation and g a field prescribed by the user. Now, these integrals are generally not sufficiently sensitive to oscillating deviations between u and u_h.

In the norm-oriented approach, the user can prescribe a semi-norm $|u - u_h|$ of the error, which will be minimized with respect to the mesh. *If the semi-norm is also a norm, then the method allows the convergence of the unknown field.* Among the possible choice of semi-norm, this can be the sum of square deviations of approximations of particular outputs. Let us take an example in aerodynamics. Let us assume that we wish to build the mesh that minimizes at the same time the errors on lift, drag and moment coefficients measured from flow solution u_h. A direct goal-oriented method will rely on a weighted sum as functional $j = \alpha_1 C_L(u) + \alpha_2 C_D(u) + \alpha_3 C_M(u)$ and minimizing the error in the evaluation of j is prone to compensation between errors committed on these three aerodynamic coefficients since when we have made $\delta j = \alpha_1(C_L(u) - C_L(u_h)) + \alpha_2(C_D(u) - C_D(u_h)) + \alpha_3(C_M(u) - C_M(u_h))$ small with an adhoc mesh, we cannot guarantee that the error on each of the three coefficients is small. A way to escape this default is to have three different functionals, with three adjoint states, to apply a necessary synthesis between the three mesh adaption solutions, which results in a notably increased computational cost. In contract, minimizing the semi-norm $|u - u_h|^2_{multi} = |C_L(u) - C_L(u_h)|^2 + |C_D(u) - C_D(u_h)|^2 + |C_M(u) - C_M(u_h)|^2$ will account for simultaneously minimizing the errors on lift, drag and moment measured from flow solution u_h with respect to mesh. Returning to the sonic boom example of section 4.6.1, good results were obtained by minimizing the error on the functional [4.18]:

$$j_{GO}(W) = \frac{1}{2} \int_\gamma \left(\frac{p - p_\infty}{p_\infty} \right)^2 \mathrm{d}\gamma.$$

Instead of minimizing with respect to the metric the error $|j_{GO}(W) - j_{GO}(W_h)|$ on this functional, it can be safer to minimize[1] the actual deviation of the pressure itself:

$$j_{NO}(\mathcal{M}) = \frac{1}{2} \int_\gamma (p_{exact} - p_h)^2 \, d\gamma.$$

In both above examples, we have chosen solely a semi-norm of the deviation to exact of the global flow field W, and therefore we do not have convergence of W when mesh is refined.

The norm-oriented method potentially addresses any kind of norm or semi-norm, with the limitation that, as for the goal-oriented method, choosing a *non-admissible* observation of the state system must be avoided. See Arian and Salas (1999) for this notion. As for the goal-oriented method, the norm-oriented method takes into account the PDE features and, contrary to the goal-oriented method, in the case where a *norm* is prescribed, it produces an approximate solution field that does converge to the exact one in this norm.

The norm-oriented method is a rather general method extending to complex CFD models, but the discussion here will be developed with the example of a 2D Poisson problem discretized by the usual linear finite-element method. This choice is motivated first by the rather complete set of theoretical works available for the finite-element approximation of a Poisson problem. This amount of theoretical background reduces as much as possible, although far from completely, the heuristics used in building the mesh adaptation analysis. A second motivation is the easy availability of exact solutions defined in a simple way. This allows us to build a kind of benchmark allowing to compare mesh adaptation methods. Let us mention also that the Poisson problem with variable coefficient is a central equation in two-fluid incompressible models as in Chapter 1 of Volume 1.

Although the usual FEM formulation relies on a projection in the H^1 norm (with first-order convergence in H^1), the user may wish to enjoy a convergence with a different norm. In this chapter, we consider the L^2 norm, which provides second-order accuracy, but the method is stated in a sufficiently general way to allow extension to many types of norms or semi-norms. The method relies on the use of a corrector field as defined in Chapter 1 of this volume and on an a priori error estimate similar to the goal-oriented method (Chapter 4 of this volume) from which is extracted the asymptotically largest terms of the local error. This allows us to minimize the norm of the synthetic error computed with the corrector. In numerical experiments, in order to evaluate the validity of the synthetic error, we compare it

---

1 The authors have not tried this option.

with the one computed with the exact solution. Although theoretical convergence results (in which functional spaces?) are not available for Euler equations, the method can be extended to this CFD model and numerical illustrations will also be given.

Section 6.2 is devoted to a reformulation of the two above-identified mesh adaptation formulations: feature-based formulation as in Chapter 5 of Volume 1, and goal-oriented formulation as in Chapter 4 of this volume. In section 6.3, we present the norm-oriented mesh adaptation method. Section 6.4 is devoted to a numerical comparison between the two field-convergent formulations, namely, Hessian-based and norm-oriented. A CFD example is then discussed.

6.2. A summary of previous analyses

The feature-based analysis of Chapter 5 of Volume 1, the error estimate and goal-oriented analysis of Chapter 4 of this volume and the corrector derivations of Chapter 1 of this volume are shortly recalled in an abstract continuous context.

6.2.1. *Feature-based adaptation by interpolation error optimization*

Let u be any sufficiently smooth function defined on Ω, which we assume in this chapter to be in $I\!\!R^2$. Let $\mathcal{M}$ be a metric of Ω. Metric $\mathcal{M}$ plays the role of the mesh and index $\mathcal{M}$ is used as indicating the approximate function $u_\mathcal{M}$ by opposition to the exact one u. We consider metrics $\mathcal{M}$ parameterizing meshes sufficiently fine for justifying the replacement of the complete approximation error by its main asymptotic part. In particular, the P^1 interpolation error $|\Pi_\mathcal{M} u - u|$ on any mesh parameterized by the metric $\mathcal{M}$ can be represented by the *continuous interpolation error* defined in Chapter 3 of Volume 1:

$$|u - \pi_\mathcal{M} u| = \frac{1}{8}\mathrm{trace}(\mathcal{M}^{-\frac{1}{2}}(\mathbf{x})|H_u(\mathbf{x})|\mathcal{M}^{-\frac{1}{2}}(\mathbf{x})),$$

where $|H_u|$ is deduced from the Hessian of u, H_u, by taking the absolute values of its eigenvalues. The functional to minimize is

$$\|u - \pi_\mathcal{M} u\|_{\mathbf{L}^p(\Omega_h)} = \left(\int_\Omega \left(\mathrm{trace}(\mathcal{M}^{-\frac{1}{2}}(\mathbf{x})|H_u(\mathbf{x})|\mathcal{M}^{-\frac{1}{2}}(\mathbf{x}))\right)^p d\mathbf{x}\right)^{\frac{1}{p}},$$

under the constraint that the complexity $\mathcal{C}(\mathcal{M})$ of the metric is equal to N. According to section 4.4 of Volume 1 (adapted to $I\!\!R^2$), we get the unique optimal $(\mathcal{M}_{\mathbf{L}^p}(\mathbf{x}))_{\mathbf{x}\in\Omega}$ as follows:

$$\mathcal{M}_{\mathbf{L}^p} = \mathcal{K}_p(H_u) \text{ with } \mathcal{K}_p(H_u) = D_{\mathbf{L}^p} \left(\det |H_u|\right)^{\frac{-1}{2p+2}} |H_u|$$
$$\text{and } D_{\mathbf{L}^p} = N \left(\int_\Omega (\det |H_u| \, \mathrm{d}\mathbf{x})^{\frac{p}{2p+2}} \right)^{-1},$$

where $D_{\mathbf{L}^p}$ is a global normalization term set to obtain a continuous metric with complexity N and $(\det |H_u|)^{\frac{-1}{2p+2}}$ is a local normalization term accounting for the sensitivity of the $\mathbf{L}^p$ norm.

6.2.2. *Implicit a priori error estimate and corrector*

We are now considering the solution of an elliptic model

$$\begin{array}{l}\text{Find } u \in V, \text{ such that } \forall\, \varphi \in V, \\ \quad a(u,\varphi) = (f,\varphi) \\ \text{with } a(v,\varphi) = \int_\Omega \frac{1}{\rho} \nabla v \cdot \nabla \varphi \, \mathrm{d}\Omega\end{array}, \qquad [6.1]$$

where ρ is a given strictly positive, bounded and possibly discontinuous field. Prescribing a complexity N of the metric $\mathcal{M}$, we consider a *still continuous* approximation of [6.1] written as follows:

$$\begin{array}{l}\text{Find } u_\mathcal{M} \in V_\mathcal{M} \subset V, \text{ such that } \forall\, \varphi \in V_\mathcal{M}, \\ \quad a(u_\mathcal{M},\varphi) = (f,\varphi).\end{array} \qquad [6.2]$$

Equation [6.2] is an abstract equation representing with smooth ingredients any approximate of [6.1] built with any unit mesh $\mathcal{H} = \mathcal{H}_\mathcal{M}$ of the metric $\mathcal{M}$:

Find $u_\mathcal{H} \in V_\mathcal{H} \subset V$, such that $\forall\, \varphi \in V_\mathcal{H}$,
$a(u_\mathcal{H},\varphi) = (f,\varphi).$

The implicit a priori error system [1.10] is then written

$$\forall\, \varphi \in V_\mathcal{H},$$
$$a(\Pi_\mathcal{H} u - u_\mathcal{H}, \varphi) = \sum_{\partial T_{ij}} \frac{1}{\rho}(\nabla\varphi|_{T_i} - \nabla\varphi|_{T_j}) \cdot \mathbf{n}_{ij} \int_{\partial T_{ij}} (\Pi_\mathcal{H} u - u) \, \mathrm{d}\sigma$$
$$- (\varphi, \Pi_\mathcal{H} f - f),$$

where T_{ij} are the triangles of a unit mesh $\mathcal{H}$ for $\mathcal{M}$.

We cannot completely transpose this system to a purely continuous one. We introduce an *a priori corrector* as introduced in section 1.2.2:

$$\bar{u}'_{prio,\mathcal{M}} \in V_{\mathcal{M}}, \quad a(\bar{u}'_{prio,\mathcal{M}}, \varphi) = K(\mathcal{M}, \varphi, u_{\mathcal{M}}) \quad \forall \varphi \in V_{\mathcal{M}} \quad \text{with}$$

$$K(\mathcal{M}, \varphi, u_{\mathcal{M}}) = \sum_{\partial T_{ij}} \frac{1}{\rho} (\nabla \varphi|_{T_i} - \nabla \varphi|_{T_j}) \cdot \mathbf{n}_{ij} \int_{\partial T_{ij}} (\pi_{\mathcal{M}} u_{\mathcal{M}} - u_{\mathcal{M}}) \, d\sigma$$

$$- (\varphi, \pi_{\mathcal{M}} f - f),$$

$$u'_{prio,\mathcal{M}} = \bar{u}'_{prio,\mathcal{M}} - (\pi_{\mathcal{M}} u_{\mathcal{M}} - u_{\mathcal{M}}).$$

6.2.3. *Goal-oriented analysis*

This section recalls in an abstract formulation the application of goal-oriented analysis to the Poisson problem as presented in Chapter 5 of this volume. We minimize the error $\delta j_{goal}(\mathcal{M})$ done when evaluating of the scalar output $j = (g, u)$. This error is simplified as follows:

$$\delta j_{goal}(\mathcal{M}) = |(g, u - u_{\mathcal{M}})| = |(g, \Pi_{\mathcal{M}} u - u_{\mathcal{M}} + u - \Pi_{\mathcal{M}} u)|.$$

Let us define the adjoint state $u^*_{g,\mathcal{M}}$:

$$\forall \psi_{\mathcal{M}} \in V_{\mathcal{M}}, \quad a(\psi_{\mathcal{M}}, u^*_{g,\mathcal{M}}) = (\psi_{\mathcal{M}}, g).$$

Then

$$(g, \Pi_{\mathcal{M}} u - u_{\mathcal{M}} + u - \Pi_{\mathcal{M}} u) = a(\Pi_{\mathcal{M}} u - u_{\mathcal{M}}, u^*_{g,\mathcal{M}}) + (g, u - \Pi_{\mathcal{M}} u)$$

and, using the analog of [1.8],

$$(g, \Pi_{\mathcal{M}} u - u_{\mathcal{M}} + u - \Pi_{\mathcal{M}} u) = a(\Pi_{\mathcal{M}} u - u, u^*_{g,\mathcal{M}}) + (f - \Pi_{\mathcal{M}} f, u^*_{g,\mathcal{M}})$$
$$- (g, \Pi_{\mathcal{M}} u - u),$$

thus

$$\delta j_{goal}(\mathcal{M}) \approx |a(\Pi_{\mathcal{M}} u - u, u^*_{g,\mathcal{M}}) + (f - \Pi_{\mathcal{M}} f, u^*_{g,\mathcal{M}}) - (g, \Pi_{\mathcal{M}} u - u)|.$$

Recall that u is unknown. The second term $\Pi_{\mathcal{M}} u - u$, similar to the main term of the Hessian-based adaptation in section 6.2.1, can be explicitly approached in the same way, that is, introducing the continuous interpolation error defined in corollary 4.1 of Chapter 4 of Volume 1. It is the same for f:

$$\delta j_{goal}(\mathcal{M}) \preceq |a(\Pi_{\mathcal{M}} u - u, u^*_{g,\mathcal{M}})| + |(f - \pi_{\mathcal{M}} f, u^*_{g,\mathcal{M}})| + |g| |\pi_{\mathcal{M}} u_{\mathcal{M}} - u_{\mathcal{M}}|.$$

For the remaining $\Pi_{\mathcal{M}} u - u$ term, we first apply lemma 5.1 and also use the continuous interpolation error $\pi_{\mathcal{M}}$. We get

$$\delta j_{goal}(\mathcal{M}) \preceq \int_{\Omega} \left(\left[\frac{1}{\rho}\rho_S(H(u^*_{g,\mathcal{M}})) + |g|\right] |\pi_{\mathcal{M}} u_{\mathcal{M}} - u_{\mathcal{M}}| + |u^*_{g,\mathcal{M}}| |\pi_{\mathcal{M}} f - f| \right) d\mathbf{x}.$$

Finally, the goal-oriented optimal metric problem is defined as follows:

$$\min_{\mathcal{M}} \int_{\Omega} \left(\left[\frac{1}{\rho}\rho_S(H(u^*_{g,\mathcal{M}})) + |g|\right] |\pi_{\mathcal{M}} u_{\mathcal{M}} - u_{\mathcal{M}}| + |u^*_{g,\mathcal{M}}| |\pi_{\mathcal{M}} f - f| \right) d\mathbf{x}.$$

When we change the parameter $\mathcal{M}$, the discrete adjoint $u^*_{g,\mathcal{M}}$ is close to its (non-zero) continuous limit and is thus not much affected, in contrast to the interpolation errors $|\pi_{\mathcal{M}} u_{\mathcal{M}} - u_{\mathcal{M}}|$ and $|\pi_{\mathcal{M}} f - f|$. We then consider, for a given $\mathcal{M}_0$, the following auxiliary optimum problem:

$$\min_{\mathcal{M}} \int_{\Omega} \left(\left[\frac{1}{\rho}\rho_S(H(u^*_{g,\mathcal{M}_0})) + |g|\right] |\pi_{\mathcal{M}} u_{\mathcal{M}} - u_{\mathcal{M}}| + |u^*_{g,\mathcal{M}_0}| |\pi_{\mathcal{M}} f - f| \right) d\mathbf{x}.$$

Expressing as in Chapter 3 of Volume 1 the abstract interpolation errors in terms of the metric $\mathcal{M}$, this will produce an optimum:

$$\mathcal{M}_{opt,\mathcal{M}_0} = Argmin_{\mathcal{M}} |tr(\mathcal{M}^{-1/2}$$
$$\times \left(\left[\frac{1}{\rho}\rho_S H(u^*_{g,\mathcal{M}_0}) + |g|\right] |H_{u_{\mathcal{M}_0}}| + |u^*_{g,\mathcal{M}_0}||H_f| \right) \mathcal{M}^{-1/2})|.$$

Observing that, in the integrand,

$$H_{goal,0} = \left[\frac{1}{\rho}\rho_S(H(u^*_{g,\mathcal{M}_0})) + |g|\right] |H_{u_{\mathcal{M}_0}}| + |u^*_{g,\mathcal{M}_0}||H_f|$$

is a positive symmetric matrix, we can apply the above calculus of variation and get

$$\mathcal{M}_{opt,\mathcal{M}_0} = \mathcal{K}_1([\frac{1}{\rho}\rho_S(H(u^*_{g,\mathcal{M}_0})) + |g|]\,|H_{u_{\mathcal{M}_0}}| + |u^*_{g,\mathcal{M}_0}|\,|H_f|),$$

where $\mathcal{K}_1$ is defined in [6.1]. This solution can then be introduced in a fixed-point loop and will produce the solution of

$$\mathcal{M}_{opt,goal} = \mathcal{K}_1([\frac{1}{\rho}\rho_S(H(u^*_{g,\mathcal{M}_{opt,goal}})) + |g|]\,|H_{u_{\mathcal{M}_{opt,goal}}}| + |u^*_{g,\mathcal{M}_{opt,goal}}|\,|H_f|).$$

6.3. Norm-oriented approach

We are now interested by the minimization of the following expression with respect to the mesh $\mathcal{M}$:

$$\delta j(\mathcal{M}) = \|u - u_\mathcal{M}\|^2_{L^2(\Omega)}. \qquad [6.3]$$

Introducing $g_\mathcal{M} = u - u_\mathcal{M}$, we get a formulation similar to the goal-oriented formulation:

$$\delta j(\mathcal{M}) = (g_\mathcal{M}, u - u_\mathcal{M}). \qquad [6.4]$$

Let us define the discrete adjoint state $u^*_\mathcal{M}$:

$$\forall \psi_\mathcal{M} \in V_\mathcal{M}, \quad a(\psi_\mathcal{M}, u^*_\mathcal{M}) = (\psi_\mathcal{M}, g_\mathcal{M}). \qquad [6.5]$$

Then, similarly to section 6.2.3, we have to solve the following optimum problem:

$$\min_\mathcal{M} \int_\Omega \left([\frac{1}{\rho}\rho_S(H(u^*_\mathcal{M})) + |g_\mathcal{M}|]\,|\pi_\mathcal{M} u_\mathcal{M} - u_\mathcal{M}| + |u^*_\mathcal{M}|\,|\pi_\mathcal{M} f - f| \right) \mathrm{d}\mathbf{x}.$$

Exactly as for section 6.2.3, we freeze the dependency of the adjoint state.

$$\min_\mathcal{M} \int_\Omega \left([\frac{1}{\rho}\rho_S(H(u^*_{\mathcal{M}_0})) + |g_\mathcal{M}|]\,|\pi_\mathcal{M} u_\mathcal{M} - u_\mathcal{M}| + |u^*_{\mathcal{M}_0}|\,|\pi_\mathcal{M} f - f| \right) \mathrm{d}\mathbf{x},$$

$$\mathcal{M}_{opt,\mathcal{M}_0} = \mathcal{K}_1\left(\left[\frac{1}{\rho}\rho_S(H(u^*_{\mathcal{M}_0}))+|g_\mathcal{M}|\right]|H_{u_{\mathcal{M}_0}}| + |u^*_{\mathcal{M}_0}||H_f|\right).$$

In this analysis, we have frozen $g_\mathcal{M} = u - u_\mathcal{M}$. This term is unknown and the norm-oriented principle is to replace it by a good and not computationally costly *corrector*. We use the corrector $u'_{prio,\mathcal{M}}$ defined in section 6.2.2. The $g_\mathcal{M}$ RHS in [6.4]–[6.5] is replaced by $u'_{prio,\mathcal{M}}$. The dependency of the corrector with respect to the mesh will be solved in the outer fixed loop as in section 6.2.3. In order to get the final norm-oriented optimum $\mathcal{M}_{opt,norm}$, we finish the computation as in Algorithm 4.1 of this volume.

The norm-oriented method presented in Algorithm 6.1. involves the solution of two extra linear systems with respect to a basic feature-based algorithm and one extra linear system with respect to a goal-oriented algorithm.

REMARK 6.1.– In contrast to the adjoint of the goal-oriented algorithm, the two auxiliary variables $u_{prio,\mathcal{M}}$ and $u^*_{prio,\mathcal{M}}$ are not consistent with a continuous adjoint and do not converge toward it when mesh is refined. They are correctors and converge to zero. A special strategy needs to be applied for solving then in a full-multigrid formulation (see Brèthes and Dervieux 2016). □

6.4. Numerical elliptic examples

The method is first demonstrated with two Poisson problems in 2D. We compare the two mesh adaptation methods that produce L^2 convergent solutions to continuous:

– the feature-based method with $p = 2$ in section 6.2.1;

– the norm-oriented method directly built on the minimization of the L^2 error norm.

We do not consider goal-oriented applications that we consider as being not convergent for the solution field.

6.4.1. *Numerical features*

In Brèthes et al. (2015), a mesh-adaptative full-multigrid (AFMG) algorithm relying on the Hessian-based adaptation criterion is proposed. We first describe in short this algorithm for the Hessian-based option. A sequence of numbers N_k of vertices is specified from a coarse mesh to finer one $N_0 = N, N_1 = 4N$,

$N_2 = 16N, N_3 = 64N, \ldots$ For each mesh size N_k, a sequence of adapted meshes of size N_k is built by iterating the following loop:

a) computing a solution;

b) computing the optimal metric;

c) building the adapted mesh.

Algorithm 6.1. Norm-oriented mesh adaptation for a steady problem

Initial mesh $\mathcal{H}_0^0$, solution W_0^0, adjoint $W_0^{*,0}$, and complexity $\mathcal{C}^0$. **While** $\alpha \leq n_{adap}$ **do**

1) Compute the discrete state $u_{\mathcal{M}^{(\alpha)}}$ on mesh $\mathcal{H}_{\mathcal{M}^{(\alpha)}}$.

2) Solve the linearized corrector system:

$$a(\bar{u}'_{prio,\mathcal{M}^{(\alpha)}}, \varphi) = \sum_{\partial T_{ij}} (\nabla\varphi|_{T_i} - \nabla\varphi|_{T_j}) \cdot \mathbf{n}_{ij} \int_{\partial T_{ij}} (\pi_{\mathcal{M}^{(\alpha)}} u_{\mathcal{M}^{(\alpha)}} - u_{\mathcal{M}^{(\alpha)}}) \, d\sigma$$
$$- (\varphi, \pi_{\mathcal{M}^{(\alpha)}} f_{\mathcal{M}^{(\alpha)}} - f_{\mathcal{M}^{(\alpha)}}), \qquad [6.6]$$

where $\pi_{\mathcal{M}^{(\alpha)}} u_{\mathcal{M}^{(\alpha)}} - u_{\mathcal{M}^{(\alpha)}}$ is expressed in terms of metric and Hessian, as in [1.11]. Put $u'_{prio,\mathcal{M}^{(\alpha)}} = \bar{u}'_{prio,\mathcal{M}^{(\alpha)}} - (\pi_{\mathcal{M}^{(\alpha)}} u_{\mathcal{M}^{(\alpha)}} - u_{\mathcal{M}^{(\alpha)}})$.

3) Then, solve the adjoint system:

$$a(\psi, u^*_{prio,\mathcal{M}^{(\alpha)}}) = (u'_{prio,\mathcal{M}^{(\alpha)}}, \psi) \qquad [6.7]$$

4) Put:

$$\mathcal{M}^{(\alpha+1)} = \mathcal{K}_1([|u'_{prio}| + \frac{1}{\rho}\rho_S H(u^*_{prio})] |H_{u_{\mathcal{M}^{(\alpha)}}}| + |u^*_{prio}||H_f|) \qquad [6.8]$$

5) Generate a new mesh $\mathcal{H}_{\mathcal{M}^{(\alpha+1)}}$ and go to 1, until convergence to a fixed point $\mathcal{M}_{opt,norm} = \mathcal{M}^{(\infty)}$.

Endwhile.

In (a), a multi-grid V-cycle is applied to a sufficient convergence. In (b), approximations of the Hessians are performed as in Chapter 5 of Volume 1. While changing the mesh, an interpolation is applied in order to enjoy a good initial condition. A prescribed number of four adaptation iterations are applied at each mesh fineness N_k.

The extension of the above loop to norm-oriented adaptation consists of replacing the single Hessian evaluation by:

– the computation of the *corrector*, using MG and, as initial solution, the previous evaluation interpolated to current mesh[2];

– the computation of the *adjoint*, using MG and, as initial solution, the previous evaluation interpolated to current mesh[3];

– the evaluation of [6.8].

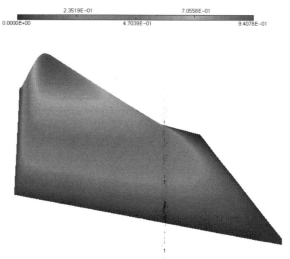

Figure 6.1. *Fully 2D boundary layer test case: sketch of the solution. For a color version of this figure, see www.iste.co.uk/dervieux/meshadaptation2*

Let us discuss computer efficiency. In the demonstrator of Brèthes et al. (2015), a particular feature is the stopping criterion of FMG that applies to the convergence of the solution of the unique solved system, that is, the system under study, $u = A^{-1}f$. It is then possible to enjoy a better initial condition and control the iterative and approximation errors convergence. Consequently, it was possible, in Brèthes et al. (2015), to show that mesh adaptation carries large improvement not only in terms of accuracy for a given number of vertices but also in terms of accuracy for a given computational time. In contrast, the method proposed in this paper involves three

---

2 The best efficiency is obtained by correcting this field as in Brèthes and Dervieux (2016).
3 The best efficiency is obtained by correcting this field as in Brèthes and Dervieux (2016).

systems to solve: (1) the system under study, $u = A^{-1}f$, (2) the corrector system and (3) the adjoint system. We shall give an idea of the performances for one test case.

6.4.2. *2D boundary layer*

This test case was proposed in Formaggia and Perotto (2003). Details of the calculation presented here can be found in Brèthes and Dervieux (2016). We solve the Poisson problem $-\Delta u = f$ in $]0,1[\times]0,1[$ with Dirichlet boundary conditions and a right-hand side f chosen for having:

$$u(x,y) = [1 - e^{-\alpha x} - (1 - e^{-\alpha})x]4y(1-y).$$

The coefficient α is chosen equal to 100. The graph of the solution is depicted in Figure 6.1. Before applying our mesh adaptative algorithm, it is interesting to evaluate the accuracy of our correctors. We choose a 161×161 uniform mesh and show in Figure 6.2 the cut of $u - u_h$ compared with the cut of both options of corrector u'. We observe that the a priori corrector does its job in a correct but inaccurate manner, while the DC one is rather accurate. We have also observed that the DC corrector does not consume notably more CPU than the a priori one. Therefore, we keep the DC option for the rest of the test case.

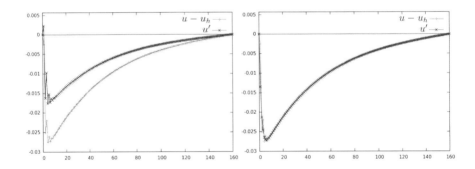

Figure 6.2. *Fully 2D boundary layer test case. Left: Comparison of the a priori corrector u'_{prio} ($\times$) with approximation error $u - u_h$ (+), error cuts for $y = 0.5$. The a priori corrector is able to correct about 60% of the approximation error amplitude. Right: Comparison of the defect-correction corrector u'_{DC} ($\times$) with the approximation error (+), error cuts for $y = 0.5$. The defect-correction corrector is able to correct about 95% of the amplitude. For a color version of this figure, see www.iste.co.uk/dervieux/meshadaptation2*

In Figure 6.3, we show a set of FMG calculations for the considered test case. The numbers of vertices of the successive meshes are supported by the horizontal axis, from 120 vertices to 30,000 vertices. The vertical axis gives the L^2-norm of the approximation error $|u - u_h|_{L^2}$ obtained on the mesh. Its variation with respect to number of vertices is compared in Figure 6.3 for the three following algorithms: (a) the uniform-mesh FMG, (b) the Hessian-based adaptive FMG and (c) the norm-oriented adaptive FMG. We observe that both adaptation methods carry an important improvement with respect to uniform-grid FMG (25,921 vertices on the finest mesh). For essentially the same number of vertices (32,318), the Hessian option gives an error divided by 47. The norm-oriented option appears as better with an error divided by 208 with 29,485 vertices. Since the exact solution u is analytically available, we could check that replacing the corrector by $u - u_h$ gives a very close convergence, in fact not better (see Brèthes and Dervieux 2016).

6.4.3. *Poisson problem with discontinuous coefficient*

This test case (also detailed in Brèthes and Dervieux 2016) exemplifies the singularity, which is met in the simulation of multi-fluid flows with a large deviation between the two values of the density, ρ_1 and ρ_2, uniform in each phase. In the case where a projection algorithm is applied to solve the Navier–Stokes equations for incompressible flow, a Poisson problem with discontinuous coefficients has to be solved as shown in section 1.3 of Volume 1. The present case does not satisfy the smoothness assumptions introduced for deriving our method, but the usual expectation in mesh adaptation is that the method also applies to non-smooth contexts. We consider the equation of Poisson $-div(\frac{1}{\rho}\nabla u) = rhs$ with a discontinuous coefficient taking two different values $1/\rho_1$ and $1/\rho_2$ on two sub-domains Ω_1 and Ω_2 separated by an interface, which is a sufficiently smooth curve for having a normal vector. This PDE is mathematically referred as a transmission problem and the solution is continuous across the interface but of discontinuous normal derivatives since $1/\rho_1 \nabla u_1 \cdot \mathbf{n} = 1/\rho_2 \nabla u_2 \cdot \mathbf{n}$, where u_1 and u_2 are the restrictions of the solution u on Ω_1 and Ω_2. In our example, we define them as $u|_{\Omega_i} = u_i = \alpha_i + \beta_i(x^2 + y^2)$ $i = 1, 2$. Further, Ω_2 is the disk of center $(0.5, 0.5)$ and radius 0.2 in the computational domain $]0,1[\times]0,1[$ and we have

$$1/\rho_1 = 1000. \ ; \ \alpha_1 = 1.23579... \ ; \ \beta_1 = -2.47158...$$
$$1/\rho_2 = 1. \quad ; \ \alpha_2 = 100. \quad ; \ \beta_2 = -2471.58... \qquad [6.9]$$

This is sketched in Figure 6.4.

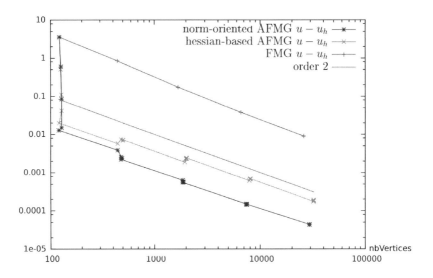

Figure 6.3. *Fully 2D boundary layer test case: convergence of the error norm $|u-u_h|_{L^2}$ as a function of number of vertices in the mesh for (+) non-adaptive FMG, (×) Hessian-based adaptive FMG and (∗) norm-oriented adaptive FMG. The quasi-vertical segments at abscissa $nbVertices \approx 128$ corresponds to the error reduction by the anisotropic mesh adaptation at constant complexity (i.e. quasi-constant number of vertices). These quasi-vertical segments corresponding to anisotropic adaptation phases also appear (less clearly) for abscissae $nbVertices \approx 512, 2048, 8196$. Oblique segments correspond to phases in which all the elements of the mesh are purely divided in four subelements by using mid-edges. As can be expected, the mesh division produces a reduction of the error by a factor of four because of second-order convergence. In contrast, anisotropic adaptation produces a tremendous error reduction for 128 vertices and not so much for the higher number of vertices. Our interpretation of this is that the main detail, the boundary layer, involves a single scale, it thickness (no smaller detail exists inside the boundary layer or elsewhere), and once the mesh has taken into account this scale, no important further improvement is possible. A last remark is that the norm-oriented approach delivers a convergence order slightly better than two while the feature-based approach delivers a slightly degraded convergence for the finer mesh, confirming that the best mesh for interpolation is not the best mesh for approximation. For a color version of this figure, see www.iste.co.uk/dervieux/meshadaptation2*

Our analysis for defining correctors uses intensively smoothness of functions and assumes, for the defect-correction option, that second-order convergence applies, which is not true for this discontinuous case. However, the injection of the correctors in the norm-oriented functional and adjoint performs adequately. The anisotropy of the finer mesh is illustrated in Figure 6.5. A convergence in terms of number of vertices with non-adaptative and Hessian-based adaptive is given in Figure 6.6. We observe that, without mesh adaptation (crosses +), the convergence order is around one. This behavior can be explained by the singularity of the solution. In contrast, the overall convergence order of the adaptative process is about two. Note that this convergence will finally deteriorate when we attain the limits of the stretching capabilities of the mesh generator. However, the second-order numerical convergence observed is a usual bonus obtained by anisotropic mesh adaptation, which has already been noted in Loseille et al. (2007). In Brèthes et al. (2015), the convergence of an anisotropic adaptation has been compared with its isotropic analogs for the same test case. The anisotropic calculation converges at order two, while the isotropic calculation converges at order $3/2$. A short analysis of these behaviors is proposed in Chapter 2 of Volume 1 and in Courty et al. (2006).

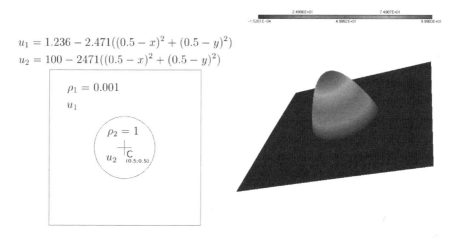

Figure 6.4. *Poisson problem with discontinuous coefficient: sketch of exact solution definition and a typical computation of it. For a color version of this figure, see www.iste.co.uk/dervieux/meshadaptation2*

Let us now compare with the feature-based method. This is the only one of our test cases for which the convergence of error in terms of number of nodes of the norm-oriented formulation is not neatly better (but it is as good) than the analog convergence of the feature-based formulation. Both adaptive algorithms give an error 10 times smaller than the non-adaptive one at a given time (Brèthes and Dervieux 2016).

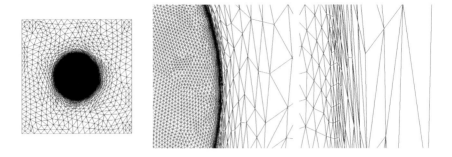

Figure 6.5. *Poisson problem with discontinuous coefficient. Views of final mesh (from left): global view, zoom on right part, zoom of the zoom*

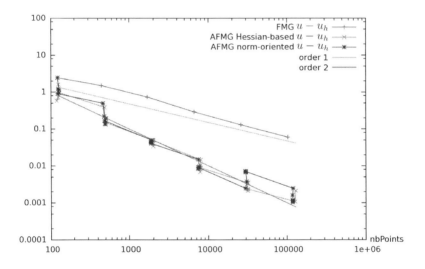

Figure 6.6. *Poisson problem with discontinuous coefficient. Convergence of the error norm $|u-u_h|_{L^2}$ as a function of number of vertices in the mesh for nonadaptative FMG (+), Hessian-based adaptative FMG (×) and norm-oriented adaptative FMG (∗). The non-adaptive convergence is solely a first-order one as expected for a singular solution. The two mesh-adaptation approaches produce second-order accuracy. Note that the mesh adaptation bonus appears in good part during the mesh enrichment phases. For a color version of this figure, see www.iste.co.uk/dervieux/meshadaptation2*

6.5. Application to flows

The above method can be extended to insviscid and viscous compressible flows. It combines the analysis of the goal-oriented approach, as defined in Chapters 4 and 5 with the correctors for Euler and Navier–Stokes computed as in Chapter 1 and introduced as g functions in the goal-oriented estimate.

6.5.1. *A comparison feature-oriented/norm*

We consider the geometry provided for the first AIAA CFD High Lift Prediction Workshop (Configuration 1). We consider an inflow at Mach 0.2 with an angle of attack of $13°$. The flow model is Euler and final rather coarse meshes are used for a more clear comparison of the three adaptation strategies which are tested: the first one controls the interpolation error on the density, velocity and pressure in L^1 norm, the second controls the interpolation error on the Mach number while the third one is based on the norm-oriented approach and controls the norm of the approximation error $||W - Wh||_{L^2}$.

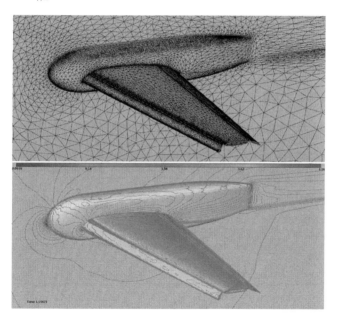

Figure 6.7. *Feature-based adaptation for minimizing the L^1 norm of the interpolation error on the density, velocity and pressure. Top: View of the skin mesh. There is not much mesh concentration on the body in the wake of wing. Bottom: Velocity on the aircraft skin. For a color version of this figure, see www.iste.co.uk/dervieux/meshadaptation2*

For each case, five adaptations at fixed complexity are performed for a total of 15 adaptations with the following complexities: [160,000, 320,000, 640,000]. This choice leads to final meshes having around one million vertices. The residual for the flow solver convergence is set to 10^{-9} for each case. The generation of the anisotropic meshes is done with the local remeshing strategy of Loseille and Menier (2013). The surface meshes and the velocity iso-lines are depicted in Figures 6.7–6.9. Depending on the adaptation strategy, rather different flow fields are observed. The adaptation on the Mach number reveals strong shear layers at the wing tip that are not present in the norm-oriented approach. On the contrary, recirculating flows are observed on the norm-oriented approach while not being observed on the Mach number adaptation. For each case, the wakes have different features. Note that the accuracy near the body is not equivalent. For the L^1 norm adaptation error and norm-oriented approaches, the far-field and inflow are much more refined than in the Mach number adaptation. This leads to better resolved phenomena for the final considered complexity. This example illustrates the need to control the whole flow field. Indeed, if the adaptation on the Mach number can provide a second-order convergent Mach number field, there is no guarantee on the other fields (density, pressure, velocity, etc.). In addition, the adaptation with the norm-oriented approach tends to increase the refinement at the inflow boundary condition and also at the far-field, although the interpolation error (on all variables) is negligible in these areas. Consequently, it seems of main interest to control all the sources of error, especially when the final intent is to certify a flow simulation.

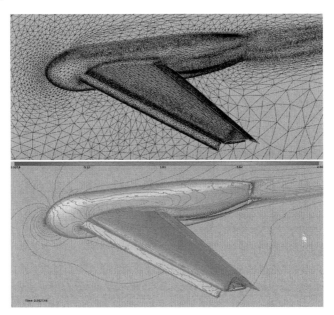

Figure 6.8. *Feature-based adaptation for minimizing the L^1 norm of the interpolation error on Mach number. Top: View of the skin mesh. There is not much mesh concentration on the body in the wake of wing. Bottom: Velocity on the aircraft skin. For a color version of this figure, see www.iste.co.uk/dervieux/meshadaptation2*

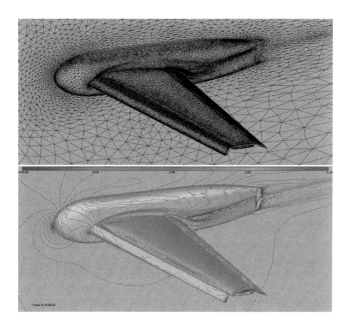

Figure 6.9. *Adaptation for minimizing the norm $||W - Wh||_{L^2}$ with the norm-oriented approach. Top: view of the skin mesh. Near-body mesh is finer and shows much more details on the aircraft body. Bottom: Velocity on the aircraft skin. For a color version of this figure, see www.iste.co.uk/dervieux/meshadaptation2*

6.5.2. *Application to a viscous flow*

The last example is the flow around a Falcon aircraft, a test case from the UMRIDA program of European Union (see Alauzet et al. (2019)). The Mach number is 0.8, the angle of attack $\alpha = 2$ degrees and a Spalart–Allmaras turbulence model is used. The norm-oriented mesh adaptation is applied, except in a region close to boundary layer. In that region, a structured boundary layer mesh is kept frozen up to $y^+ = 500$, while the mesh is solely adapted in the upper boundary layer region and the outer field. We choose to control the interpolation error on the local Mach number in L^2 norm. Fifteen mesh adaptation iterations are performed. We split the adaptation loop into three phases with an increasing theoretical complexity (outside of the boundary layer region). Within each step, the adapted mesh at a fixed theoretical complexity is converged in five iterations. The final adapted meshes for each step contain 2,298,958, 6,168,815 and 10,337,483 vertices for theoretical complexities of, respectively, 100,000, 200,000 and 400,000. The final adapted mesh (for the largest complexity) is illustrated in Figure 6.10.

130 Mesh Adaptation for Computational Fluid Dynamics 2

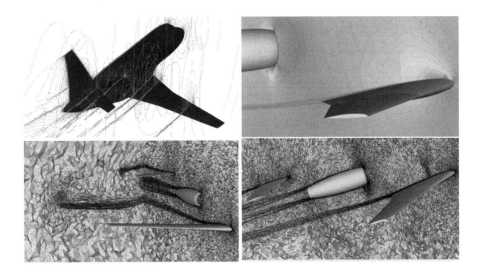

Figure 6.10. *Top: Mach solution field. Bottom: final adapted mesh. Shape: courtesy of Dassault Aviation. For a color version of this figure, see www.iste.co.uk/dervieux/meshadaptation2*

Such adapted meshes considerably enhance the efficiency of the flow solver and the solution accuracy. We first note that the wake is highly resolved and the wing tip vortices are well captured. Second, mesh refinements along the shock on the upper surface of the wing lead to an accurate computation of the shock – boundary layer interaction. We also observe a nice transition between the boundary layer structured mesh and the adapted anisotropic mesh.

6.6. Conclusion

The norm-oriented mesh adaptation method gives an answer to a well-formulated problem. Considering a numerical scheme and prescribing a norm, we want to find the mesh giving the smallest approximation error in that norm for a given number of vertices. The norm-oriented mesh adaptation method transforms the problem into an optimization problem, which is mathematically well-posed. It relies on the following other features.

A corrector represents the approximation error at the cost of a linearized state. In Chapter 1 of this volume, we give two examples of correctors for an elliptic model. An a priori corrector is built from the variational discrete statement. A defect-correction corrector is built from a finer-mesh defect correction principle. These correctors appear as not always very accurate but sufficiently accurate for our purpose. According to the type of approximation, at least the second one, defect-correction is extendable to many models and schemes.

The norm-oriented algorithm is presented as a natural extension of the goal-oriented algorithm which, in our formulation, is itself an extension of the Hessian-based algorithm. More precisely, while the Hessian-based algorithm solves only the PDE under study in the mesh-adaptation loop, the goal-oriented algorithm also solves an adjoint system (with linearized operator, transposed). The norm-oriented algorithm solves three systems, a corrector (linearized system with an adhoc RHS), an adjoint (linearized and transposed with the corrector as RHS) and the PDE itself. The three algorithms have in common an anisotropic a priori error analysis and a metric-based mesh parameterization. The feature-based method produces convergent solution fields but does not take into account the precise equation and discretization. The goal-oriented method takes into account equation and discretization but is too focused on a particular output and does not generally produce convergent solution fields. The norm-oriented method has the advantages of both.

6.7. Notes

Some other numerical examples are presented in Brèthes and Dervieux (2016).

Other approaches. Taking into account the influence of the PDE on the error through an *equation-based estimate* has been an important topic in the literature. The formulation of *goal-oriented methods* was an important step for a more justified error evaluation. It has been introduced in Becker and Rannacher (1996). It relied on an a posteriori estimate. A good synthesis concerning a posteriori estimates is given in Verfürth (2013); see also Ern and Vohralík (2015). An interest of a posteriori estimate is that it is expressed in terms of the approximate solution, assumed to be available in a mesh adaptation loop. A second interest is that it does not require the use of higher order (approximate) derivatives, in contrast to truncation analyses. These estimates show accurately where the mesh should be refined. A method for deducing a better anisotropic mesh from an a posteriori estimate, such as the one from Becker and Rannacher (1996), is proposed in Formaggia and Perotto (2003), while a theory for H^p norms in Agouzal and Vassilevski (2010) and a joint analysis of H^p and L^p norms of the error are presented in Agouzal et al. (2010). These methods cannot focus on an arbitrary user-specified error norm but relies on a particular one, specified by the variational formulation of the PDE. A more popular option is to choose, as accuracy target, a particular scalar output depending on the PDE solution. Any scalar output can be considered, except that difficulties can arise for the so-called non-admissible ones, according to Arian and Salas (1999). An a posteriori estimate also allows for building *correctors* for goal-oriented analyses (Giles and Pierce 1999; Pierce and Giles 2000). In Venditti and Darmofal (2000), the goal-oriented approach is cleverly combined with the correction strategy of Pierce and Giles (2000) and with the Hessian-based metric approach, still minimizing the interpolation error of a user-prescribed feature.

7

Goal-Oriented Adaptation for Unsteady Flows

Chapter 2 of this volume presents a transient fixed-point (TFP) multi-scale algorithm for the adaptation of mesh to an unsteady flow. Chapter 4 of this volume presents a fully anisotropic goal-oriented (GO) mesh adaptation technique for a steady flow, which uses an adjoint state and allows to define the best mesh for the most accurate evaluation of a given scalar output. The present chapter addresses the extension to unsteady flows of the GO method. Its rationale is to combine the GO of Chapter 4 of this volume with the TFP advances of Chapter 2 of this volume. However, the resulting GO-TFP has to be *global in time*, that is, the whole time interval needs to be computed before improved meshes are derived. Examples of applications concerns unsteady Euler flows.

7.1. Introduction

In this chapter, we define an extension of the feature-based TFP introduced in Chapter 2 of this volume to a GO formulation. To this end, several methodological issues need to be addressed. First, an error analysis based on the so-called state and adjoint systems is developed in order to set an unsteady mesh optimization problem. Second, as a necessary adaptation of the method introduced in Chapter 2, we define a global transient fixed-point (GTFP) algorithm for solving the coupled system formed, this time of three fields, the unsteady state, the unsteady adjoint state and the adapted meshes. By "global" we mean that mesh is not adapted successively time interval after time interval, but after the computation of the whole physical time of the simulation. This algorithm will be analyzed a priori and its convergence rate to the continuous solution optimized. Third, at the computer algorithmic level, it is also necessary to master the computational (memory and time) cost of the new system, which couples a time-forward state, a time-backward adjoint and a mesh update based on metric statistics over the time interval.

This chapter starts with a formal description (section 7.2) of the unsteady error analysis in its most general expression, then (section 7.3) the application to unsteady compressible Euler flows is presented. In section 7.4, the optimal adjoint-based metric is defined and all its relative issues, then section 7.5 is dedicated to strategies for mesh convergence and algorithm optimization. In section 7.6, the GTFP mesh adaptation algorithm is presented. This chapter is concluded with numerical illustrations with blast wave and acoustic wave propagation.

7.2. Formal error analysis

Let us introduce an unsteady system of PDEs in its variational formulation:

$$\text{Find } w \in \mathcal{V} \text{ such that } \forall \varphi \in \mathcal{V}, \quad (\Psi(w), \varphi) = 0. \tag{7.1}$$

Here, the space $\mathcal{V}$ is made up of *space–time functions*, being a set of sufficiently smooth functions defined on $.\Omega \times]0, T[$, Ω being a spatial domain. The associated discrete variational formulation is written as

$$\text{Find } w_h \in \mathcal{V}_h \text{ such that } \forall \varphi_h \in \mathcal{V}_h, \quad (\Psi_h(w_h), \varphi_h) = 0, \tag{7.2}$$

where $\mathcal{V}_h$ is a subspace of $\mathcal{V}$. For a solution w of state system [7.1], we define a *space–time functional output* as

$$j(w) \in \mathbb{R} \,;\, j = (g, w), \tag{7.3}$$

where (g, w) holds for the following rather general functional output formulation:

$$(g, w) = \int_0^T \int_\Omega (g_\Omega, w) \, d\Omega \, dt + \int_\Omega (g_T, w(T)) \, d\Omega + \int_0^T \int_\Gamma (g_\Gamma, w) \, d\Gamma \, dt, \tag{7.4}$$

where g_Ω, g_T and g_Γ are assumed to be sufficiently regular functions. We introduce the *continuous adjoint* w^*, solution of the following system:

$$w^* \in \mathcal{V}, \, \forall \psi \in \mathcal{V}, \, \left(\frac{\partial \Psi}{\partial w}(w)\psi, w^* \right) = (g, \psi). \tag{7.5}$$

The objective here is to estimate the following approximation error committed on the functional

$$\delta j = j(w) - j(w_h),$$

where w and w_h are, respectively, solutions of [7.1] and [7.2]. Using the fact that $\mathcal{V}_h \subset \mathcal{V}$, the following error estimates for the unknown can be written as

$$(\Psi_h(w), \varphi_h) - (\Psi_h(w_h), \varphi_h) = (\Psi_h(w), \varphi_h) - (\Psi(w), \varphi_h) = ((\Psi_h - \Psi)(w), \varphi_h). \quad [7.6]$$

It is then useful to choose the test function φ_h as the discrete adjoint state, $\varphi_h = w_h^*$, which is the solution of

$$\forall \psi_h \in \mathcal{V}_h, \; \left(\frac{\partial \Psi_h}{\partial w_h}(w_h)\psi_h, w_h^* \right) = (g, \psi_h). \quad [7.7]$$

We assume that w_h^* is close to the continuous adjoint state w^*. The same formal development as in Chapter 4 of this volume leads to the following *a priori* formal estimate:

$$\delta j \approx ((\Psi_h - \Psi)(w), w^*). \quad [7.8]$$

The next section is devoted to the application of estimator [7.8] to the unsteady Euler model.

7.3. Unsteady Euler models

7.3.1. *Continuous state system and finite volume formulation*

7.3.1.1. Continuous state system

The 3D unsteady compressible Euler equations are set in the computational space–time domain $\mathcal{Q} = \Omega \times [0, T]$, where T is the (positive) maximal time and $\Omega \subset \mathbb{R}^3$ is the spatial domain. In order to easily use the usual local interpolator Π_h, we define our working functional space as $V = \left[H^1(\Omega) \cap \mathcal{C}(\bar{\Omega}) \right]^5$, that is the set of measurable functions that are continuous with square integrable gradient. We formulate the Euler model in a compact manner with a variational formulation in the functional space $\mathcal{V} = H^1\{[0,T]; V\}$:

Find $W \in \mathcal{V}$ such that $\forall \varphi \in \mathcal{V}$, $(\Psi(W), \varphi) = 0$

with $(\Psi(W), \varphi) = \int_\Omega \varphi(0)(W_0 - W(0)) \, d\Omega + \int_0^T \int_\Omega \varphi W_t \, d\Omega \, dt$

$+ \int_0^T \int_\Omega \varphi \nabla \cdot \mathcal{F}(W) \, d\Omega \, dt - \int_0^T \int_\Gamma \varphi \hat{\mathcal{F}}(W).\mathbf{n} \, d\Gamma \, dt.$ [7.9]

Here, W is the vector of conservative flow variables and $\mathcal{F}(W) = (\mathcal{F}_1(W), \mathcal{F}_2(W), \mathcal{F}_3(W))$ is the usual Euler flux (see [1.1] and [1.2] in Volume 1). Functions φ and W have five components, and therefore the product φW holds for $\sum_{k=1..5} \varphi_k W_k$. We have denoted by Γ the boundary of the computational domain Ω, $\mathbf{n}$ is the outward normal to Γ, $W(0)(\mathbf{x}) = W(\mathbf{x}, t)|_{t=0}$ for any $\mathbf{x}$ in Ω, W_0 the initial condition and the boundary flux $\hat{\mathcal{F}}$ contains the different boundary conditions, which involve inflow, outflow and slip boundary conditions.

7.3.1.2. *Discrete state system*

We first consider a spatially semi-discrete model, derived from Chapter 1 of Volume 1. We reformulate it under the form of a finite element variational formulation, this time in the unsteady context. We assume that Ω is covered by a finite-element partition in simplicial elements denoted as K. The mesh, denoted by $\mathcal{H}$, is the set of the elements. Let us introduce the following approximation space:

$V_h = \left\{ \varphi_h \in V \mid \varphi_{h|K} \text{ is affine } \forall K \in \mathcal{H} \right\}$, and $\mathcal{V}_h = H^1\{[0,T]; V_h\} \subset \mathcal{V}.$

Let Π_h be the usual $\mathcal{P}^1$ projector:

$\Pi_h : V \to V_h$ such that $\Pi_h \varphi(\mathbf{x}_i) = \varphi(\mathbf{x}_i)$, $\forall \mathbf{x}_i$ vertex of $\mathcal{H}$.

We extend it to time-dependent functions:

$\Pi_h : H^1\{[0,T]; V\} \to \mathcal{V}_h$ such that $(\Pi_h \varphi)(t) = \Pi_h(\varphi(t))$, $\forall t \in [0,T]$.

The weak discrete formulation is written as follows:

Find $W_h \in \mathcal{V}_h$ such that $\forall \varphi_h \in \mathcal{V}_h$, $(\Psi_h(W_h), \varphi_h) = 0$, with:

$$(\Psi_h(W_h), \varphi_h) = \int_\Omega \varphi_h(0)(\Pi_h W_h(0) - W_{0h}) \, d\Omega$$
$$+ \int_0^T \int_\Omega \varphi_h \, \Pi_h W_{h,t} \, d\Omega \, dt + \int_0^T \int_\Omega \varphi_h \nabla \cdot \mathcal{F}_h(W_h) d\Omega dt$$
$$- \int_0^T \int_\Gamma \varphi_h \hat{\mathcal{F}}_h(W_h).\mathbf{n} d\Gamma dt + \int_0^T \int_\Omega \varphi_h \, D_h(W_h) \, d\Omega \, dt,$$

with $\mathcal{F}_h = \Pi_h \mathcal{F}$ and $\hat{\mathcal{F}}_h = \Pi_h \hat{\mathcal{F}}$. The D_h term accounts for the numerical diffusion. It involves the difference between the Galerkin central-differences approximation and the second-order Godunov approximation of Chapter 1 of Volume 1. Far from singularities, the D_h term is a third-order term with respect to the mesh size, which we do not take into account in the adaptation analysis. For shocked fields, monotonicity limiters produce first-order terms, but as in previous chapters, we let the formally second-order Hessian estimate concentrate nodes in the shocks. As concerns time discretization, we consider the following explicit RK1 time integration:

$$\Psi_h^n(W_h^n, W_h^{n-1}) = \frac{W_h^n - W_h^{n-1}}{\delta t^n} + \Phi_h(W_h^{n-1}) = 0 \quad \text{for } n = 1, ..., N. \qquad [7.10]$$

The time-dependent functional is discretized as follows:

$$j_h(W_h) = \sum_{n=1}^N \delta t^n j_h^{n-1}(W_h^{n-1}).$$

7.3.2. *Continuous adjoint system and discretization*

7.3.2.1. *Continuous adjoint system*

We refer here to the continuous adjoint system [7.5] introduced previously:

$$W^* \in \mathcal{V}, \; \forall \psi \in \mathcal{V} : \; \left(\frac{\partial \Psi}{\partial W}(W)\psi, W^*\right) - (g, \psi) = 0. \qquad [7.11]$$

We recall that (g, ψ) is defined by [7.4]. Replacing $\Psi(W)$ by its definition [7.9] and integrating by parts, we get

$$\left(\frac{\partial \Psi}{\partial W}(W)\psi, W^*\right) = \int_\Omega (\psi(0)W^*(0) - \psi(T)W^*(T))\, d\Omega$$

$$+ \int_0^T \int_\Omega \psi \left(-W_t^* - \left(\frac{\partial \mathcal{F}}{\partial W}\right)^* \nabla W^*\right) d\Omega\, dt \qquad [7.12]$$

$$+ \int_0^T \int_\Gamma \psi \left[\left(\frac{\partial \mathcal{F}}{\partial W}\right)^* W^*.\mathbf{n} - \left(\frac{\partial \hat{\mathcal{F}}}{\partial W}\right)^* W^*.\mathbf{n}\right] d\Gamma\, dt.$$

Consequently, the continuous adjoint state W^* must be such that

$$-W_t^* - \left(\frac{\partial \mathcal{F}}{\partial W}\right)^* \nabla W^* = g_\Omega \quad \text{in } \Omega, \qquad [7.13]$$

with the associated adjoint boundary conditions

$$\left(\frac{\partial \mathcal{F}}{\partial W}\right)^* W^*.\mathbf{n} - \left(\frac{\partial \hat{\mathcal{F}}}{\partial W}\right)^* W^*.\mathbf{n} = g_\Gamma \quad \text{on } \Gamma$$

and the adjoint state final condition

$$W^*(T) = g_T.$$

The adjoint Euler equations is a system of advection equations, where the temporal integration goes backwards, that is, in the opposite direction of usual time. Thus, when solving the unsteady adjoint system, we start at the end of the flow run and progress back until reaching the start time.

7.3.2.2. *Discrete adjoint system*

Although any consistent approximation of the continuous adjoint system could be built by discretizing [7.13], we choose the option to build the discrete adjoint system from the discrete state system defined by [7.10], [7.10] in order to be closer to the true error from which the continuous model is derived. For the sake of simplicity, we restrict to the case $g_T = 0$ for the functional output defined by [7.4]. The problem of minimizing the error committed on the target functional $j(W_h) = (g, W_h)$, subject to the Euler system [7.10], can be transformed into an unconstrained problem for the following Lagrangian functional:

$$\mathcal{L}(W_h, W_h^*) = \sum_{n=1}^{N} \delta t^n j_h^{n-1}(W_h^{n-1}) - \sum_{n=1}^{N} \delta t^n (W_h^{*,n})^T \Psi_h^n(W_h^n, W_h^{n-1}),$$

where $W_h^{*,n}$ are the N vectors of the Lagrange multipliers (which are the time-dependent adjoint states). The conditions for an extremum are

$$\frac{\partial \mathcal{L}}{\partial W_h^{*,n}} = 0 \quad \text{and} \quad \frac{\partial \mathcal{L}}{\partial W_h^n} = 0, \quad \text{for } n = 1, ..., N.$$

The first condition is clearly verified from [7.10]. Thus, the Lagrangian multipliers $W_h^{*,n}$ must be chosen such that the second condition of extrema is verified. This provides the unsteady discrete adjoint system

$$\begin{cases} W_h^{*,N} = 0 \\ W_h^{*,n-1} = W_h^{*,n} + \delta t^n \dfrac{\partial j_h^{n-1}}{\partial W_h^{n-1}}(W_h^{n-1}) - \delta t^n (W_h^{*,n})^T \dfrac{\partial \Phi_h}{\partial W_h^{n-1}}(W_h^{n-1}), \end{cases} \quad [7.14]$$

or equivalently, the semi-discrete unsteady adjoint model reads

$$\Psi_h^{*,n}(W_h^{*,n}, W_h^{*,n-1}, W_h^{n-1}) = \frac{W_h^{*,n-1} - W_h^{*,n}}{-\delta t^n} + \Phi_h^*(W_h^{*,n}, W_h^{n-1}) = 0$$

for $n = 1, ..., N$

with

$$\Phi_h^*(W_h^{*,n}, W_h^{n-1}) = \frac{\partial j_h^{n-1}}{\partial W_h^{n-1}}(W_h^{n-1}) - (W_h^{*,n})^T \frac{\partial \Phi_h}{\partial W_h^{n-1}}(W_h^{n-1}).$$

As the adjoint system runs in reverse time, the first expression in the adjoint system [7.14] is referred to as adjoint "initialization".

7.3.2.3. *The memory requirement issue*

Computing $W_h^{*,n-1}$ at time t^{n-1} requires the knowledge of state W_h^{n-1} and adjoint state $W_h^{*,n}$. Therefore, the knowledge of all states $\{W_h^{n-1}\}_{n=1,N}$ is needed to compute backward the adjoint state from time T to 0 (Figure 7.1), which involves large memory storage effort. For instance, if we consider a 3D simulation with a mesh composed of 1 million vertices, then we need to store at each iteration 5 millions solution data (we have five conservative variables). If we perform 1,000 iterations, then the memory effort to store all states is 37.25 Gb for double-type data storage (or 18.62 for float-type data storage). Two strategies are employed to reduce importantly this drawback: *checkpointing* and *interpolation*. The memory effort can be reduced by out-of-core storage of *checkpoints* as shown in Figure 7.2. First the state-simulation is performed to store checkpoints. Second, when computing backward the adjoint, we first recompute all states from the checkpoint and store them in memory and then we compute the unsteady adjoint until the checkpoint physical time. Therefore, with this method, the state W storage only concerns checkpoints and the timesteps inside the current checkpoint interval, with the cost of a double computation. The second strategy consists of storing solution states in memory only every m solver iterations. When the unsteady adjoint is solved, solution states between two savings are linearly *interpolated*. This second method leads to a loss of accuracy for the unsteady adjoint computation. We then adopt and recommend the checkpointing option.

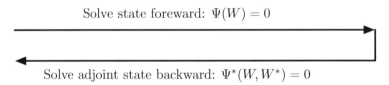

Figure 7.1. *Memory issue for computing an unsteay adjoint. When state equation, going forward in time, is nonlinear, the adjoint equation, going backward in time cannot be solved without the knowledge of the state W*

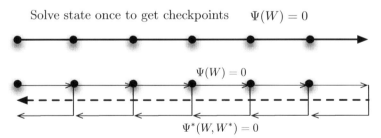

Figure 7.2. *The memory issue for computing an unsteady adjoint is solved by checkpointing. On each checkpoint interval, state W is forward recomputed from a stored value at beginning of the interval. These recomputed values permit the backward advancing of the adjoint through the checkpoint interval*

7.3.3. *Impact of the adjoint: numerical example*

Before going more deeply in the error model, we would like to emphasize how strongly the use of an adjoint may impact the density distribution of adapted meshes. The simulation of a blast in a 2D geometry representing a city is performed (see Figure 7.3). A blast-like initialization $W_{blast} = (10, 0, 0, 250)$ in ambient air $W_{air} = (1, 0, 0, 2.5)$ is considered in a small region of the computational domain. We perform a forward/backward computation on a uniform mesh of 22,574 vertices and 44,415 triangles. Output functional of interest j is the quadratic deviation from ambient pressure on target surface S, which is a part of the higher building roof (Figure 7.3):

$$j(W) = \int_0^T \int_S \frac{1}{2}(p(t) - p_{air})^2 \, dS \, dt.$$

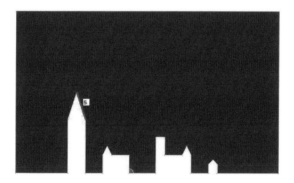

Figure 7.3. *Initial blast solution (about center of bottom) and location of target surface S. For a color version of this figure, see www.iste.co.uk/dervieux/meshadaptation2*

Figure 7.4 plots the density isolines of the flow at different times showing several shock waves traveling throughout the computational domain. Figure 7.5 depicts the associated density adjoint state progressing backward in time. The same a-dimensional physical time is considered for both figures.

The simulation points out the ability of the adjoint to automatically provide the sensitivity of the flow field on the functional. Early in the simulation (top left picture), many information are *captured where adjoint is non-zero*. We notice that shock waves which will directly impact the targeted surface are clearly detected by the adjoint, but also shocks waves reflected by the left building which will be redirected toward surface S. At the middle of the simulation, the adjoint neglects waves that are traveling in the direction opposite to S and also waves that will not impact surface S before final time T since they will not have an influence on the cost functional. While getting closer to final time T (bottom right picture), the adjoint only focuses on the last waves that will impact surface S and ignores the rest of the flow.

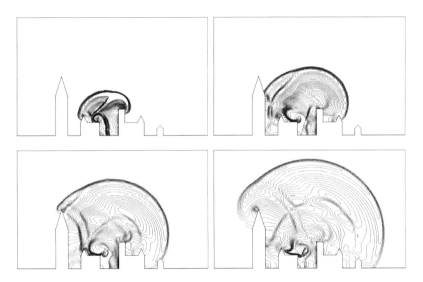

Figure 7.4. *2D city blast solution state evolution. From left to right and top to bottom, snapshot of the density isolines at a-dimensional time 1.2, 2.25, 3.3 and 4.35*

7.4. Optimal unsteady adjoint-based metric

7.4.1. *Error analysis for the unsteady Euler model*

In estimation [7.8], operators Ψ and Ψ_h are now replaced by their expressions given by relations [7.9] and [7.10]. In Chapter 4 of this volume, dealing with the steady case, it is observed that even for shocked flows, it is interesting to neglect the numerical viscosity term. We follow again this option. We also discard for simplicity the error committed when imposing approximatively the initial condition. We finally get the following error model:

$$\delta j \approx \int_0^T \int_\Omega W^* \left(W - \Pi_h W\right)_t \mathrm{d}\Omega \, \mathrm{d}t + \int_0^T \int_\Omega W^* \nabla.(\mathcal{F}(W)$$
$$- \Pi_h \mathcal{F}(W)) \, \mathrm{d}\Omega \, \mathrm{d}t - \int_0^T \int_\Gamma W^* \left(\hat{\mathcal{F}}(W) - \Pi_h \hat{\mathcal{F}}(W)\right).\mathbf{n} \, \mathrm{d}\Gamma \, \mathrm{d}t. \quad [7.15]$$

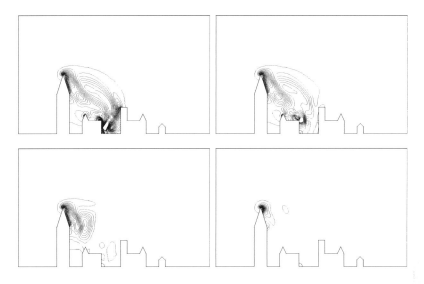

Figure 7.5. *2D city blast adjoint state evolution. From left to right and top to bottom, snapshot of the adjoint-density isolines at a-dimensional time 1.2, 2.25, 3.3 and 4.35*

Integrating by parts leads to

$$\delta j \approx \int_0^T \int_\Omega W^* \left(W - \Pi_h W\right)_t \mathrm{d}\Omega\, \mathrm{d}t - \int_0^T \int_\Omega \nabla W^* \left(\mathcal{F}(W)\right.$$
$$\left. - \Pi_h \mathcal{F}(W)\right) \mathrm{d}\Omega\, \mathrm{d}t - \int_0^T \int_\Gamma W^* \left(\bar{\mathcal{F}}(W) - \Pi_h \bar{\mathcal{F}}(W)\right).\mathbf{n}\, \mathrm{d}\Gamma\, \mathrm{d}t. \quad [7.16]$$

with $\bar{\mathcal{F}} = \hat{\mathcal{F}} - \mathcal{F}$. We observe that this estimate of δj is expressed in terms of interpolation errors of the Euler fluxes and of the time derivative weighted by continuous functions W^* and ∇W^*.

7.4.1.1. *Error bound with a safety principle*

The integrands in error estimation [7.16] contain positive and negative parts, which can compensate for some particular meshes. It is preferred here to not rely on these possibly parasitic effects and to (hopefully slightly) overestimate the error[1]. To this end, all integrands are bounded by their absolute values:

---

1 A comparison of both standpoints is given in Chapter 5 of this volume.

$$(g, W_h - W) \leq \int_0^T \int_\Omega |W^*| \, |(W - \Pi_h W)_t| \, d\Omega \, dt + \int_0^T \int_\Omega |\nabla W^*| \, |\mathcal{F}(W)$$
$$- \Pi_h \mathcal{F}(W)| \, d\Omega \, dt + \int_0^T \int_\Gamma |W^*| \, |(\bar{\mathcal{F}}(W) - \Pi_h \bar{\mathcal{F}}(W)) . \mathbf{n}| \, d\Gamma \, dt. \qquad [7.17]$$

7.4.2. *Continuous error model*

According to the continuous mesh framework, introduced in Chapter 3, a continuous mesh of computational domain Ω is identified to a Riemannian metric field $\mathcal{M} = (\mathcal{M}(\mathbf{x}))_{\mathbf{x} \in \Omega}$ and the diagonalization of $\mathcal{M}(\mathbf{x})$:

$$\mathcal{M}(\mathbf{x}) = d^{\frac{2}{3}}(\mathbf{x}) \, \mathcal{R}(\mathbf{x}) \begin{pmatrix} r_1^{-\frac{2}{3}}(\mathbf{x}) & & \\ & r_2^{-\frac{2}{3}}(\mathbf{x}) & \\ & & r_3^{-\frac{2}{3}}(\mathbf{x}) \end{pmatrix} {}^t \mathcal{R}(\mathbf{x}), \qquad [7.18]$$

with same notations as in Chapter 3. Given a smooth function u, to each unit mesh $\mathcal{H}$ with respect to $\mathbf{M}$ corresponds a local interpolation error $|u - \Pi u|$. In section 4.2 of Chapter 3 of Volume 1, it is shown that all these interpolation errors are well represented by the so-called continuous interpolation error related to $\mathbf{M}$, which is expressed locally in terms of the Hessian H_u of u as follows:

$$(u - \pi_\mathcal{M} u)(\mathbf{x}, t) = \frac{1}{10} \text{trace}(\mathcal{M}^{-\frac{1}{2}}(\mathbf{x}) \, |H_u(\mathbf{x}, t)| \, \mathcal{M}^{-\frac{1}{2}}(\mathbf{x})), \qquad [7.19]$$

where $|H_u|$ is deduced from H_u by taking the absolute values of its eigenvalues and where time-dependency notations have been added for use in next sections. Working in this framework enables us to write estimate [7.17] in a continuous form:

$$|(g, W_h - W)| \approx \mathbf{E}(\mathbf{M}) = \int_0^T \int_\Omega |W^*| \, |(W - \pi_\mathcal{M} W)_t| \, d\Omega \, dt$$
$$+ \int_0^T \int_\Omega |\nabla W^*| \, |\mathcal{F}(W) - \pi_\mathcal{M} \mathcal{F}(W)| \, d\Omega \, dt$$
$$+ \int_0^T \int_\Gamma |W^*| \, |(\bar{\mathcal{F}}(W) - \pi_\mathcal{M} \bar{\mathcal{F}}(W)) . \mathbf{n}| \, d\Gamma \, dt. \qquad [7.20]$$

We observe that the third term introduce a dependency of the error with respect to the boundary surface mesh. As in Chapter 4 of this volume, we discard this term.

Then, introducing the continuous interpolation error, we can write the *simplified error model* as follows:

$$\mathbf{E}(\mathbf{M}) = \int_0^T \int_\Omega \text{trace}\left(\mathcal{M}^{-\frac{1}{2}}(\mathbf{x},t)\,\mathbf{H}(\mathbf{x},t)\,\mathcal{M}^{-\frac{1}{2}}(\mathbf{x},t)\right) d\Omega\, dt \qquad [7.21]$$

with $\mathbf{H}(\mathbf{x},t) = \sum_{j=1}^{5} \left([\Delta t]_j(\mathbf{x},t) + [\Delta x]_j(\mathbf{x},t) + [\Delta y]_j(\mathbf{x},t) + [\Delta z]_j(\mathbf{x},t)\right),$

in which

$$[\Delta t]_j(\mathbf{x},t) = \left|W_j^*(\mathbf{x},t)\right| \cdot \left|H(W_{j,t})(\mathbf{x},t)\right|, [\Delta x]_j(\mathbf{x},t)$$
$$= \left|\frac{\partial W_j^*}{\partial x}(\mathbf{x},t)\right| \cdot \left|H(\mathcal{F}_1(W_j))(\mathbf{x},t)\right|, \qquad [7.22]$$

$$[\Delta y]_j(\mathbf{x},t) = \left|\frac{\partial W_j^*}{\partial y}(\mathbf{x},t)\right| \cdot \left|H(\mathcal{F}_2(W_j))(\mathbf{x},t)\right|, [\Delta z]_j(\mathbf{x},t)$$
$$= \left|\frac{\partial W_j^*}{\partial z}(\mathbf{x},t)\right| \cdot \left|H(\mathcal{F}_3(W_j))(\mathbf{x},t)\right|. \qquad [7.23]$$

Here, W_j^* denotes the jth component of the adjoint vector W^*, $H(\mathcal{F}_i(W_j))$ is the Hessian of the jth component of the vector $\mathcal{F}_i(W)$ and $H(W_{j,t})$ is the Hessian of the jth component of the time derivative of W. The mesh optimization problem is written as follows:

$$\text{Find } \mathbf{M}_{opt} = \text{Argmin}_\mathbf{M}\, \mathbf{E}(\mathbf{M}), \qquad [7.24]$$

under the *space-time* complexity constraint:

$$\mathcal{C}_{st}(\mathbf{M}) = N_{st}, \qquad [7.25]$$

where N_{st} is the *total space–time (st) prescribed complexity*, modeling the total space–time number of nodes. In this unsteady context, the *total space–time (st) complexity* $\mathcal{C}_{st}(\mathbf{M})$ is a time integral of spatial mesh complexity weighted by the inverse time step:

$$\mathcal{C}_{st}(\mathbf{M}) = \int_0^T \tau(t)^{-1} \left(\int_\Omega d_\mathcal{M}(\mathbf{x},t)d\mathbf{x}\right) dt \qquad [7.26]$$

where $\tau(t)$ is the time step used at time t of interval $[0,T]$.

7.4.3. *Spatial minimization for a fixed* t

Let us assume that at time t, we seek for the optimal continuous mesh $\mathbf{M}_{go}(t)$ that minimizes the instantaneous error, that is, the spatial error for a fixed time t:

$$\tilde{\mathbf{E}}(\mathbf{M}(t)) = \int_\Omega \text{trace}\left(\mathcal{M}^{-\frac{1}{2}}(\mathbf{x},t)\,\mathbf{H}(\mathbf{x},t)\,\mathcal{M}^{-\frac{1}{2}}(\mathbf{x},t)\right)\,\mathrm{d}\mathbf{x}$$

under the constraint that the number of vertices is prescribed to

$$\mathcal{C}(\mathbf{M}(t)) = \int_\Omega d_{\mathcal{M}(t)}(\mathbf{x},t)\,\mathrm{d}\mathbf{x} = N(t). \qquad [7.27]$$

Similarly to section 4.4 of Chapter 4 of this volume, solving the optimality conditions provides the *optimal goal-oriented ("GO") instantaneous continuous mesh* $\mathbf{M}_{go}(t) = (\mathcal{M}_{go}(\mathbf{x},t))_{\mathbf{x}\in\Omega}$ at time t defined by

$$\mathcal{M}_{go}(\mathbf{x},t) = N(t)^{\frac{2}{3}}\,\mathcal{M}_{go,1}(\mathbf{x},t), \qquad [7.28]$$

where $\mathcal{M}_{go,1}$ is the optimum for $\mathcal{C}(\mathbf{M}(t)) = 1$:

$$\mathcal{M}_{go,1}(\mathbf{x},t) = \left(\int_\Omega (\det \mathbf{H}(\bar{\mathbf{x}},t))^{\frac{1}{5}}\,\mathrm{d}\bar{\mathbf{x}}\right)^{-\frac{2}{3}} (\det \mathbf{H}(\mathbf{x},t))^{-\frac{1}{5}}\,\mathbf{H}(\mathbf{x},t). \qquad [7.29]$$

The corresponding optimal instantaneous error at time t is written as:

$$\tilde{\mathbf{E}}(\mathbf{M}_{go}(t)) = 3\,N(t)^{-\frac{2}{3}}\left(\int_\Omega (\det \mathbf{H}(\mathbf{x},t))^{\frac{1}{5}}\,\mathrm{d}\mathbf{x}\right)^{\frac{5}{3}} = 3\,N(t)^{-\frac{2}{3}}\,\mathcal{K}(t). \qquad [7.30]$$

For the sequel of this chapter, we denote: $\mathcal{K}(t) = \left(\int_\Omega (\det \mathbf{H}(\mathbf{x},t))^{\frac{1}{5}}\,\mathrm{d}\mathbf{x}\right)^{\frac{5}{3}}$.

7.4.4. *Temporal minimization*

To complete the resolution of optimization problem [7.24–7.25], we perform a temporal minimization in order to get the optimal space–time continuous mesh. In other words, we need to find the optimal time law $t \mapsto N(t)$ for the instantaneous mesh size. First, we consider the simpler case where the time step τ is specified by the user as a function of time $t \mapsto \tau(t)$. Second, we deal with the case of an explicit time advancing solver subject to Courant time step condition.

7.4.4.1. *Temporal minimization for specified τ*

Let us consider the case where the time step τ is specified by a function of time $t \mapsto \tau(t)$. After the spatial optimization, the space–time error is written as

$$\mathbf{E}(\mathbf{M}_{go}) = \int_0^T \tilde{\mathbf{E}}(\mathbf{M}_{go}(t))\, dt = 3 \int_0^T N(t)^{-\frac{2}{3}} \mathcal{K}(t)\, dt, \qquad [7.31]$$

and we aim at minimizing it under the following space–time complexity constraint:

$$\int_0^T N(t)\tau(t)^{-1}\, dt = N_{st}. \qquad [7.32]$$

In other words, we concentrate on seeking for *the optimal distribution of $N(t)$ when the space–time total number of nodes N_{st} is prescribed*. Let us apply the one-to-one change in variables:

$$\tilde{N}(t) = N(t)\tau(t)^{-1} \quad \text{and} \quad \tilde{\mathcal{K}}(t) = \tau(t)^{-\frac{2}{3}} \mathcal{K}(t).$$

Then, our temporal optimization problem becomes

$$\min_{\mathbf{M}} \mathbf{E}(\mathbf{M}) = \int_0^T \tilde{N}(t)^{-\frac{2}{3}} \tilde{\mathcal{K}}(t)\, dt \quad \text{under constraint} \quad \int_0^T \tilde{N}(t)\, dt = N_{st}.$$

The solution of this problem is given by

$$\tilde{N}_{opt}(t)^{-\frac{5}{3}} \tilde{\mathcal{K}}(t) = const \quad \Rightarrow \quad N_{opt}(t) = C(N_{st})\, (\tau(t) \mathcal{K}(t))^{\frac{3}{5}}.$$

Here, the constant $C(N_{st})$ can be obtained by introducing the above expression in space–time complexity constraint [7.32], leading to

$$C(N_{st}) = \left(\int_0^T \tau(t)^{-\frac{2}{5}} \mathcal{K}(t)^{\frac{3}{5}}\, dt \right)^{-1} N_{st},$$

which completes the description of the optimal space–time metric for a prescribed time step. Using relation [7.28], the analytic expression of the optimal space–time GO metric $\mathbf{M}_{go}$ is written as

$$\mathcal{M}_{go}(\mathbf{x},t) = N_{st}^{\frac{2}{3}} \left(\int_0^T \tau(t)^{-\frac{2}{5}} \left(\int_\Omega (\det \mathbf{H}(\bar{\mathbf{x}},t))^{\frac{1}{5}} d\bar{\mathbf{x}} \right) dt \right)^{-\frac{2}{3}}$$

$$\times \tau(t)^{\frac{2}{5}} (\det \mathbf{H}(\mathbf{x},t))^{-\frac{1}{5}} \mathbf{H}(\mathbf{x},t). \qquad [7.33]$$

We get the following optimal error:

LEMMA 7.1.–

$$\mathbf{E}(\mathbf{M}_{go}) = 3\, N_{st}^{-\frac{2}{3}} \left(\int_0^T \tau(t)^{-\frac{2}{5}} \left(\int_\Omega (\det \mathbf{H}(\mathbf{x},t))^{\frac{1}{5}} d\mathbf{x} \right) dt \right)^{\frac{5}{3}}. \; \square \qquad [7.34]$$

7.4.4.2. *Temporal minimization for explicit time advancing*

In the case of an explicit time advancing subject to a Courant condition, we get a more complex context, since time step strongly depends on the smallest mesh size. We restrict to the case of smooth data and solution. We still seek for the optimal continuous mesh that minimizes space–time error [7.31] under complexity constraint [7.32]. Let $\Delta x_{min,1}(t) = \min_\mathbf{x} \min_i h_i(\mathbf{x})$ be the smallest mesh size of $\mathbf{M}_{go,1}(t)$. Since the metric definition [7.18] is homogeneous with the inverse square of mesh size, we deduce the smallest mesh size of $\mathbf{M}_{go}(t)$ from [7.28]:

$$\Delta x_{min}(t) = N(t)^{-\frac{1}{3}} \Delta x_{min,1}(t),$$

where $\Delta x_{min,1}(t)$ is independent of the mesh complexity. A way to write the Courant condition for time advancing is to define the time step $\tau(t)$ by

$$\tau(t) = c(t)^{-1} \Delta x_{min}(t) = N(t)^{-\frac{1}{3}} c(t)^{-1} \Delta x_{min,1}(t),$$

where $c(t)$ is the maximal wave speed over the domain at time t. Again, we search for the optimal distribution of $N(t)$ when the space–time complexity N_{st} is prescribed (see equation [7.32]), with

$$N_{st} = \int_0^T N(t)^{\frac{4}{3}} c(t) (\Delta x_{min,1}(t))^{-1} dt.$$

We choose to apply the one-to-one change in variables:

$$\hat{N}(t) = N(t)^{\frac{4}{3}} c(t) \left(\Delta x_{min,1}(t)\right)^{-1} \quad \text{and} \quad \hat{\mathcal{K}}(t) = \mathcal{K}(t) c(t)^{\frac{1}{2}} \left(\Delta x_{min,1}(t)\right)^{-\frac{1}{2}}.$$

Therefore, the corresponding space–time approximation error over the simulation time interval and space–time complexity reduces to

$$\mathbf{E}(\mathbf{M}_{go}) = 3 \int_0^T N(t)^{-\frac{2}{3}} \mathcal{K}(t) \, dt = 3 \int_0^T \hat{N}(t)^{-\frac{1}{2}} \hat{\mathcal{K}}(t) \, dt$$

and $\quad \int_0^T \hat{N}(t) dt = N_{st}.$

This optimization problem has for its solution:

$$\hat{N}_{opt}(t)^{-\frac{3}{2}} \hat{\mathcal{K}}(t) = const \ \Rightarrow \ \hat{N}_{opt}(t) = C(N_{st}) \, \hat{\mathcal{K}}(t)^{\frac{2}{3}},$$

and by considering the space–time complexity constraint relation we deduce

$$C(N_{st}) = N_{st} \left(\int_0^T \hat{\mathcal{K}}(t)^{\frac{2}{3}} \, dt \right)^{-1}.$$

Using the definitions of $\hat{N}$ and $\hat{\mathcal{K}}$ in the above relations, we get

$$N(t)^{\frac{4}{3}} c(t) (\Delta x_{min,1}(t))^{-1} = N_{st} \left(\int_0^T \left(\mathcal{K}(t) c(t)^{\frac{1}{2}} (\Delta x_{min,1}(t))^{-\frac{1}{2}} \right)^{\frac{2}{3}} dt \right)^{-1}$$

$$\times \left(\mathcal{K}(t) \, c(t)^{\frac{1}{2}} \, (\Delta x_{min,1}(t))^{-\frac{1}{2}} \right)^{\frac{2}{3}}$$

$$\iff N(t) = N_{st}^{\frac{3}{4}} c(t)^{-\frac{1}{2}} (\Delta x_{min,1}(t))^{\frac{1}{2}} \mathcal{K}(t)^{\frac{1}{2}}$$

$$\left(\int_0^T c(t)^{\frac{1}{3}} (\Delta x_{min,1}(t))^{-\frac{1}{3}} \mathcal{K}(t)^{\frac{2}{3}} \, dt \right)^{-\frac{3}{4}}.$$

Consequently, after some simplifications, we obtain the following expression of the optimal space–time GO continuous mesh $\mathbf{M}_{go}$ and error:

LEMMA 7.2.–

$$\mathcal{M}_{go}(\mathbf{x},t) = N_{st}^{\frac{1}{2}} \left(\int_0^T \theta(t)^{\frac{1}{3}} \mathcal{K}(t)^{\frac{2}{3}} \, dt \right)^{-\frac{1}{2}}$$

$$\times \theta(t)^{-\frac{1}{3}} \mathcal{K}(t)^{-\frac{1}{15}} (\det \mathbf{H}(\mathbf{x},t))^{-\frac{1}{5}} \mathbf{H}(\mathbf{x},t) \qquad [7.35]$$

$$\mathbf{E}(\mathbf{M}_{go}) = 3 N_{st}^{-\frac{1}{2}} \left(\int_0^T \theta(t)^{\frac{1}{3}} \mathcal{K}(t)^{\frac{2}{3}} \, dt \right)^{\frac{3}{2}}, \qquad [7.36]$$

where $\theta(t) = c(t) \left(\Delta x_{min,1}(t) \right)^{-1}$ and $\mathcal{K}(t) = \left(\int_\Omega (\det \mathbf{H}(\mathbf{x},t))^{\frac{1}{5}} \, d\mathbf{x} \right)^{\frac{5}{3}}$. □

7.4.5. *Temporal minimization for time sub-intervals*

The previous analysis provides the optimal size of the adapted meshes for each computational time level. Hence, this analysis requires the mesh to be adapted at each flow solver time step. But, in practice this approach involves a very large number of remeshing, which is CPU consuming and spoils solution accuracy due to many solution transfers from one mesh to a new mesh. We adopt the strategy introduced in Chapter 2 of this volume where the number of remeshings is controlled (thus drastically reduced) by generating adapted meshes used for several time steps. We split the simulation time interval into n_{adap} sub-intervals $[t_{i-1}, t_i]$ for $i = 1, .., n_{adap}$. Each spatial mesh $\mathbf{M}^i$ is then kept constant during all the timesteps of each sub-interval $[t_{i-1}, t_i]$. We could consider this partition as a *time discretization of the mesh adaptation problem*. In order to evaluate the number of nodes N^i of the ith adapted mesh $\mathbf{M}^i$ on sub-interval $[t_{i-1}, t_i]$ we could take a mean of the time-continuous analog, that is, inspired by [7.26], we could define

$$N^i = \frac{\int_{t_{i-1}}^{t_i} N_{opt}(t) \tau(t)^{-1} dt}{\int_{t_{i-1}}^{t_i} \tau(t)^{-1} dt}.$$

This is a – maybe not optimal – projection of the optimal time-continuous context. Here, we propose a different option in which we get an optimal discrete answer.

7.4.5.1. *Spatial minimization on a sub-interval*

Given the continuous mesh complexity N^i for the single adapted mesh used during time sub-interval $[t_{i-1}, t_i]$, we seek for the optimal continuous mesh $\mathbf{M}^i_{go}$ solution of the following problem:

$$\min_{\mathbf{M}^i} \mathbf{E}^i(\mathbf{M}^i) = \int_\Omega \text{trace}\left((\mathcal{M}^i)^{-\frac{1}{2}}(\mathbf{x}) \mathbf{H}^i(\mathbf{x}) (\mathcal{M}^i)^{-\frac{1}{2}}(\mathbf{x})\right) d\mathbf{x}$$

such that $\mathcal{C}(\mathbf{M}^i) = N^i,$

where matrix $\mathbf{H}^i$ on the sub-interval can be defined by either using an $\mathbf{L}^1$ or an $\mathbf{L}^\infty$ norm:

$$\mathbf{H}^i_{\mathbf{L}^1}(\mathbf{x}) = \int_{t_{i-1}}^{t_i} \mathbf{H}(\mathbf{x}, t)\, dt \quad \text{or} \quad \mathbf{H}^i_{\mathbf{L}^\infty}(\mathbf{x}) = (t_i - t_{i-1}) \max_{t \in [t_{i-1}, t_i]} \mathbf{H}(\mathbf{x}, t).$$

Processing as previously, we get the following:

LEMMA 7.3.– (Spatial optimality condition on a time interval)

$$\mathcal{M}^i_{go}(\mathbf{x}) = (N^i)^{\frac{2}{3}} \mathcal{M}^i_{go,1}(\mathbf{x}) \quad \text{with } \mathcal{M}^i_{go,1}(\mathbf{x})$$
$$= \left(\int_\Omega (\det \mathbf{H}^i(\bar{\mathbf{x}}))^{\frac{1}{5}} d\bar{\mathbf{x}}\right)^{-\frac{2}{3}} (\det \mathbf{H}^i(\mathbf{x}))^{-\frac{1}{5}} \mathbf{H}^i(\mathbf{x}).$$

The corresponding optimal error $\mathbf{E}^i(\mathbf{M}^i_{go})$ is written as

$$\mathbf{E}^i(\mathbf{M}^i_{go}) = 3\,(N^i)^{-\frac{2}{3}} \left(\int_\Omega (\det \mathbf{H}^i(\mathbf{x}))^{\frac{1}{5}} d\mathbf{x}\right)^{\frac{5}{3}} = 3\,(N^i)^{-\frac{2}{3}} \mathcal{K}^i. \; \square$$

To complete our analysis, we shall perform a temporal minimization. Again, we first consider the case where the time step τ is specified by a function of time and, second, we deal then with the case of an explicit time advancing solver subject to Courant time step condition.

7.4.5.2. *Temporal minimization for specified τ*

After the spatial minimization, the temporal optimization problem reads

$$\min_{\mathbf{M}} \mathbf{E}(\mathbf{M}) = \sum_{i=1}^{n_{adap}} \mathbf{E}^i(\mathbf{M}_{go}^i) = 3 \sum_{i=1}^{n_{adap}} (N^i)^{-\frac{2}{3}} \mathcal{K}^i$$

$$\text{such that} \quad \sum_{i=1}^{n_{adap}} N^i \left(\int_{t_{i-1}}^{t_i} \tau(t)^{-1} \mathrm{d}t \right) = N_{st}.$$

We set the one-to-one mapping:

$$\tilde{N}^i = N^i \left(\int_{t_{i-1}}^{t_i} \tau(t)^{-1} \mathrm{d}t \right) \quad \text{and} \quad \tilde{\mathcal{K}}^i = \mathcal{K}^i \left(\int_{t_{i-1}}^{t_i} \tau(t)^{-1} \mathrm{d}t \right)^{\frac{2}{3}},$$

then the optimization problem reduces to

$$\min_{\mathbf{M}} \sum_{i=1}^{n_{adap}} (\tilde{N}^i)^{-\frac{2}{3}} \tilde{\mathcal{K}}^i \quad \text{such that} \quad \sum_{i=1}^{n_{adap}} \tilde{N}^i = N_{st}.$$

The solution is

$$\tilde{N}_{opt}^i = \mathcal{C}(N_{st}) \, (\tilde{\mathcal{K}}^i)^{\frac{3}{5}} \quad \text{with} \quad \mathcal{C}(N_{st}) = N_{st} \left(\sum_{i=1}^{n_{adap}} (\tilde{\mathcal{K}}^i)^{\frac{3}{5}} \right)^{-1}$$

$$\Rightarrow N^i = N_{st} \left(\sum_{i=1}^{n_{adap}} (\mathcal{K}^i)^{\frac{3}{5}} \left(\int_{t_{i-1}}^{t_i} \tau(t)^{-1} \mathrm{d}t \right)^{\frac{2}{5}} \right)^{-1} (\mathcal{K}^i)^{\frac{3}{5}} \left(\int_{t_{i-1}}^{t_i} \tau(t)^{-1} \mathrm{d}t \right)^{-\frac{3}{5}}.$$

and we deduce the following:

LEMMA 7.4.– *The optimal continuous mesh* $\mathbf{M}_{go} = \{\mathbf{M}_{go}^i\}_{i=1,\ldots,n_{adap}}$ *and error are given by*

$$\mathcal{M}_{go}^i(\mathbf{x}) = N_{st}^{\frac{2}{3}} \left(\sum_{i=1}^{n_{adap}} (\mathcal{K}^i)^{\frac{3}{5}} \left(\int_{t_{i-1}}^{t_i} \tau(t)^{-1} \mathrm{d}t \right)^{\frac{2}{5}} \right)^{-\frac{2}{3}} \left(\int_{t_{i-1}}^{t_i} \tau(t)^{-1} \mathrm{d}t \right)^{-\frac{2}{5}}$$

$$\times (\det \mathbf{H}^i(\mathbf{x}))^{-\frac{1}{5}} \mathbf{H}^i(\mathbf{x}) \qquad [7.37]$$

$$\mathbf{E}(\mathbf{M}_{go}) = 3\, N_{st}^{-\frac{2}{3}} \left(\sum_{i=1}^{n_{adap}} (\mathcal{K}^i)^{\frac{3}{5}} \left(\int_{t_{i-1}}^{t_i} \tau(t)^{-1} \mathrm{d}t \right)^{\frac{2}{5}} \right)^{\frac{5}{3}}, \qquad [7.38]$$

with $(\mathcal{K}^i)^{\frac{3}{5}} = \int_\Omega (\det \mathbf{H}^i(\mathbf{x}))^{\frac{1}{5}} \mathrm{d}\mathbf{x}$. □

7.4.5.3. *Temporal minimization for explicit time advancing*

Taking into account the variable time step controlled by a CFL condition changes the terms of the optimal mesh problem. Similarly to the previous section, the Courant-based time-step is written as

$$\tau(t) = c(t)^{-1} \Delta x_{min}^i = (N^i)^{-\frac{1}{3}} c(t)^{-1} \Delta x_{min,1}^i \quad \text{for } t \in [t_{i-1}, t_i],$$

where $\Delta x_{min,1}^i$ is the smallest altitude of $\mathbf{M}_{go,1}^i$ and $c(t)$ is the maximal wave speed over the domain. The optimization problem is written as

$$\min_{\mathbf{M}} \mathbf{E}(\mathbf{M}) = \sum_{i=1}^{n_{adap}} \mathbf{E}^i(\mathbf{M}_{go}^i) = 3 \sum_{i=1}^{n_{adap}} (N^i)^{-\frac{2}{3}} \mathcal{K}^i$$

under the constraint

$$\sum_{i=1}^{n_{adap}} (N^i)^{\frac{4}{3}} \left(\int_{t_{i-1}}^{t_i} c(t) (\Delta x_{min,1}^i)^{-1} \mathrm{d}t \right) = N_{st}.$$

We set again:

$$\hat{N}^i = (N^i)^{\frac{4}{3}} \left(\int_{t_{i-1}}^{t_i} c(t) (\Delta x_{min,1}^i)^{-1} \mathrm{d}t \right)$$

$$\text{and} \quad \hat{\mathcal{K}}^i = \mathcal{K}^i \left(\int_{t_{i-1}}^{t_i} c(t) (\Delta x_{min,1}^i)^{-1} \mathrm{d}t \right)^{\frac{1}{2}}.$$

Then, the optimization problem becomes

$$\min_{\mathbf{M}} \mathbf{E}(\mathbf{M}) = \sum_{i=1}^{n_{adap}} \mathbf{E}^i(\mathbf{M}_{go}^i) = 3 \sum_{i=1}^{n_{adap}} (N^i)^{-\frac{2}{3}} \mathcal{K}^i = 3 \sum_{i=1}^{n_{adap}} (\hat{N}^i)^{-\frac{1}{2}} \hat{\mathcal{K}}^i$$

under the constraint

$$\sum_{i=1}^{n_{adap}} (N^i)^{\frac{4}{3}} \left(\int_{t_{i-1}}^{t_i} c(t) \, (\Delta x^i_{min,1})^{-1} \mathrm{d}t \right) = \sum_{i=1}^{n_{adap}} \hat{N}^i = N_{st}.$$

This optimization problem has for its solution

$$\hat{N}^i_{opt} = \mathcal{C}(N_{st}) \, (\hat{\mathcal{K}}^i)^{\frac{2}{3}} \quad \text{with} \quad \mathcal{C}(N_{st}) = N_{st} \left(\sum_{i=1}^{n_{adap}} (\hat{\mathcal{K}}^i)^{\frac{2}{3}} \right)^{-1},$$

from which we deduce

$$N^i_{opt} = N_{st}^{\frac{3}{4}} \left(\sum_{i=1}^{n_{adap}} (\mathcal{K}^i)^{\frac{2}{3}} \left(\int_{t_{i-1}}^{t_i} c(t) \, (\Delta x^i_{min,1})^{-1} \mathrm{d}t \right)^{\frac{1}{3}} \right)^{-\frac{3}{4}}$$

$$\times (\mathcal{K}^i)^{\frac{1}{2}} \left(\int_{t_{i-1}}^{t_i} c(t) \, (\Delta x^i_{min,1})^{-1} \mathrm{d}t \right)^{-\frac{1}{2}}.$$

For the sake of clarity, we set $\theta^i = \int_{t_{i-1}}^{t_i} c(t) \, (\Delta x^i_{min,1})^{-1} \mathrm{d}t$.

LEMMA 7.5.– The optimal continuous mesh $\mathbf{M}_{go} = \{\mathbf{M}^i_{go}\}_{i=1,\dots,n_{adap}}$ and error reads

$$\mathcal{M}^i_{go}(\mathbf{x}) = N_{st}^{\frac{1}{2}} \left(\sum_{i=1}^{n_{adap}} (\mathcal{K}^i)^{\frac{2}{3}} (\theta^i)^{\frac{1}{3}} \right)^{-\frac{1}{2}}$$

$$(\theta^i)^{-\frac{1}{3}} (\mathcal{K}^i)^{-\frac{1}{15}} (\det \mathbf{H}^i(\mathbf{x}))^{-\frac{1}{5}} \mathbf{H}^i(\mathbf{x}) \qquad [7.39]$$

$$\mathbf{E}(\mathbf{M}_{go}) = 3 \, N_{st}^{-\frac{1}{2}} \left(\sum_{i=1}^{n_{adap}} (\mathcal{K}^i)^{\frac{2}{3}} (\theta^i)^{\frac{1}{3}} \right)^{\frac{3}{2}} . \quad \square \qquad [7.40]$$

REMARK 7.1.– The choice of the time sub-intervals $\{[t_{i-1}, t_i]\}_{i=1, n_{adap}}$ for a given n_{adap} is a mesh adaptation problem: What is the subdivision of interval $[0, T]$ in n_{adap} sub-intervals, which will minimize the error on optimal mesh/metric $\mathcal{M}$? Since we take a constant metric in the sub-interval, the error main term in approximating $\mathcal{M}$ is the following integral of the absolute value of the time derivative of $\mathcal{M}$:

$$\sum_{i=1}^{n_{adap}} \int_{t_{i-1}}^{t_i} \left| \frac{\partial \mathcal{M}(t)}{\partial t} \right| \mathrm{d}t,$$

which can be minimized by equidistribution of the error by sub-interval. □

REMARK 7.2.– The parameter n_{adap}, number of time sub-intervals, is important for the efficiency of the adaptation algorithm. When a too large value is prescribed for n_{adap}, the error in mesh-to-mesh interpolation may increase. A compromise needs to be found by the user.

7.5. Theoretical mesh convergence analysis

Our motivation in developing mesh adaptation algorithms is to get approximation algorithms with better convergence to continuous target data. By better, we mean that we want ideally:

i) to reach the asymptotic high-order convergence of scheme;

ii) with a lower number of nodes;

iii) even for solutions involving discontinuities.

These properties have been previously obtained in the context of *steady* flows in Chapters 6 of Volume 1 and Chapters 4 and 5 of this volume. In the present chapter property (i), order two is not obtained, as discussed in the sequel[2]. In the following, the convergence order of numerical solutions computed with the presented adaptive platform is established for all the different strategies described above. First, the case of smooth flows is given, then the case of singular flows is exemplified by the case of a traveling discontinuity.

7.5.1. *Smooth flow fields*

For some Hessian-based methods, like the $\mathbf{L}^\infty - \mathbf{L}^p$ approach of Chapter 2 of this volume, an analysis can be proposed for predicting the order of convergence to the continuous solution. According to remark 4.1, any generic GO method does not provide flow field convergence. In other words, for the present algorithm, the adaptive state solution W_h generally does *not* converge to the continuous one W. On

---

[2] The cure which we propose is to introduce a multirate time-advancing like the one described in Chapter 3 of this volume.

the contrary, in the case of regular solutions, the expression of the optimal error model indicates that the functional output indeed converges, and with a predictable order.

7.5.1.1. *Smooth context with specified time step*

Here, we are adapting the mesh with the purpose of reducing the spatial error, with a time-step specified once for all. The usual spatial analysis extends and gives in 3D

$$\mathbf{E}(\mathbf{M}_{go}) = O(N_{st}^{-\frac{2}{3}}) \quad \text{as} \quad N_{st} \to \infty,$$

for the case of an adaptation at each time step [7.34] and also for the case of an adaptation with a fixed mesh at each sub-interval [7.38].

LEMMA 7.6.– For smooth solution, the modeled *spatial* error on functional for a specified time-step converges at *second-order rate*.

7.5.1.2. *Smooth context with Courant-based time step*

According to the Appendix of Chapter 2 of this volume, we are adapting the mesh $\mathbf{M}(t)$ in order to, by the magic of Courant condition, reduce both space and time error. The space–time dimension is 4. Now, in this case, we have established the following estimate:

$$\mathbf{E}(\mathbf{M}_{go}) = O(N_{st}^{-\frac{1}{2}}) \quad \text{as} \quad N_{st} \to \infty,$$

for the case of an adaptation at each time step [7.36], and also for the case of an adaptation with a fixed mesh at each sub-interval [7.40].

LEMMA 7.7.– For smooth solution, the *space–time* modeled error on functional for Courant-based time step converges at *second-order rate*.

7.5.1.3. *Singular flow fields*

An important advantage of mesh adaptation is its potential ability in addressing with a better mesh convergence the case of solution fields involving (isolated) singularities such as discontinuities, etc. In this case, it can happen that the specified mesh density becomes locally infinitely large and the mesh size can be zero. In the unsteady case, the time step becomes also zero or too small. Not only the time advancing is stopped or too slow, but also the evaluation of the global effort (defined by the mesh size divided by the inverse time step) becomes difficult to use. In practice this can be avoided by forcing the mesh size to be larger than a prescribed length. Adaptative mesh convergence can be preserved if this safety parameter is adequately decreased when global complexity is increased. According to Chapter 2 of Volume 1, higher order space convergence can be obtained *also* for singular case with discontinuities, assuming as usually that the error on singularity is concentrated on a subset of zero measure of the computational domain to be discretized by the mesh. However, even in that case, the convergence barrier lemma 2.9 in Volume 1 gives a space–time convergence order bounded by $8/5$.

7.6. From theory to practice

In practice, it remains to approximate the three-field coupled system

$$W \in \mathcal{V}, \forall \varphi \in \mathcal{V}, (\Psi(\mathcal{M}, W), \varphi) = 0 \qquad \text{"Euler system"}$$

$$W^* \in \mathcal{V}, \forall \psi \in \mathcal{V}, \left(\frac{\partial \Psi}{\partial W}(\mathcal{M}, W)\psi, W^*\right) = (g, \psi) \qquad \text{"adjoint system"}$$

$$\mathcal{M}(\mathbf{x}, t) = \mathcal{M}_{go}(\mathbf{x}, t) \qquad \text{"adapted mesh"}$$

$$[7.41]$$

by a discrete one and then solve it. For discretizing the state and adjoint PDEs, we take the spatial schemes introduced previously that are explicit Runge–Kutta time-advancing schemes. Such time discretization methods have nonlinear stability properties like TVD, which are particularly suitable for the integration of system of hyperbolic conservation laws where discontinuities appear. Discretizing the last equation consists of specifying the mesh according to a discrete metric deduced from the discrete states. In Chapter 2 of Volume 1, we describe a TFP that avoids the generation of a new mesh at each solver iteration, which otherwise would imply serious degradation of the CPU time and the solution accuracy due to the large number of mesh modifications. It is also an answer to the *lag problem* occurring when, from the solution at t^n, a new adapted mesh is generated at level t^n to compute next solution at level t^{n+1}. The basic idea consists of splitting the simulation time frame $[0, T]$ into n_{adap} adaptation sub-intervals:

$$[0, T] = [0 = t_0, t_1] \cup \ldots \cup [t_i, t_{i+1}] \cup \ldots \cup [t_{n_{adap}-1}, t_{n_{adap}}],$$

and to keep the same adapted mesh for each sub-interval. Consequently, the time-dependent simulation is performed with only n_{adap} different adapted meshes. The mesh used on each sub-interval is adapted to control the solution accuracy from t_{i-1} to t_i. We examine now how to apply this program when a GO approach is chosen.

7.6.1. *Choice of the GO metric*

The optimal adapted meshes for each sub-interval are generated according to analysis of section 7.4.5. In this work, the following particular choice has been made:

– the Hessian metric for sub-interval i is based on a control of the temporal error in $\mathbf{L}^\infty$ norm:

$$\mathbf{H}^i_{\mathbf{L}^\infty}(\mathbf{x}) = \Delta t_i \max_{t \in [t_i, t_{i+1}]} \mathbf{H}(\mathbf{x}, t) = \Delta t_i \, \mathbf{H}^i_{\max}(\mathbf{x}) ;$$

– the function $\tau : t \to \tau(t)$ is constant and equal to 1;

– all sub-intervals have the same time length Δt.

The optimal GO metric $\mathbf{M}_{go} = \{\mathbf{M}^i_{go}\}_{i=1,\ldots,n_{adap}}$ then simplifies to

$$\mathcal{M}^i_{go}(\mathbf{x}) = N_{st}^{\frac{2}{3}} \left(\sum_{i=1}^{n_{adap}} (\int_\Omega (\det \mathbf{H}^i_{\max}(\mathbf{x}))^{\frac{1}{5}} d\mathbf{x}) \right)^{-\frac{2}{3}} (\Delta t)^{\frac{1}{3}}$$

$$\left(\det \mathbf{H}^i_{\max}(\mathbf{x})\right)^{-\frac{1}{5}} \mathbf{H}^i_{\max}(\mathbf{x}).$$

REMARK 7.3.– We note that we obtain a similar expression of the optimal metric to the one proposed in Alauzet and Olivier (2011), but it is presently obtained in a GO context and by means of a space–time error minimization.

7.6.2. *Global fixed-point mesh adaptation algorithm*

To converge the nonlinear mesh adaptation problem, that is, converging the couple mesh solution, we propose a fixed-point mesh adaptation algorithm. The parameter N_{st} representing the total computational effort is prescribed by the user and will influence the size of all the meshes defined during the time interval. That is, to compute any metric fields $\mathbf{M}^i_{go}$, we have to evaluate a global normalization factor that requires the knowledge of all $\mathbf{H}^i_{\max}$. Thus, the whole simulation from 0 to T must be performed to be able to evaluate all metrics $\mathbf{M}^i_{go}$.

Note that passing to a global – over $[0, T]$ – adaptation iteration is solely an *option* when a feature-based adaptation is chosen (e.g. Alauzet and Olivier 2011). In contrast, choosing the *GTFP mesh adaptation algorithm* (Algorithm 7.1) is compulsory for the GO approach.

Algorithm 7.1. Global transient fixed-point mesh adaptation algorithm

–//– Fixed-point loop to converge global space-time mesh adaptation

For j=1,n_{nptfx}

//– Solve state once to get checkpoints
For i=1,n_{adap}

- $\mathcal{W}_{0,i}^j = \texttt{ConservativeSolutionTransfer}(\mathcal{H}_{i-1}^j, \mathcal{W}_{i-1}^j, \mathcal{H}_i^j)$

- $\mathcal{W}_i^j = \texttt{SolveStateForward}(\mathcal{W}_{0,i}^j, \mathcal{H}_i^j)$

End for

//– Solve state and adjoint backward and store samples
For i=n_{adap}, 1

- $(\mathcal{W}^*)_i^j = \texttt{AdjointStateTransfer}(\mathcal{H}_{i+1}^j, (\mathcal{W}_0^*)_{i+1}^j, \mathcal{H}_i^j)$

- $\{\mathcal{W}_i^j(k), (\mathcal{W}^*)_i^j(k)\}_{k=1,n_k} = \texttt{SolveStateAndAdjointBackward}(\mathcal{W}_{0,i}^j, (\mathcal{W}^*)_i^j, \mathcal{H}_i^j)$

- $|\mathbf{H}_{\max}|_i^j = \texttt{ComputeGoalOrientedHessianMetric}(\mathcal{H}_i^j, \{\mathcal{W}_i^j(k), (\mathcal{W}^*)_i^j(k)\}_{k=1,n_k})$

End for

- $\mathcal{C}^j = \texttt{ComputeSpaceTimeComplexity}(\{|\mathbf{H}_{\max}|_i^j\}_{i=1,n_{adap}})$

- $\{\mathcal{M}_i^j\}_{i=1,n_{adap}} = \texttt{ComputeUnsteadyGoalOrientedMetrics}(\mathcal{C}^j, \{|\mathbf{H}_{\max}|_i^j\}_{i=1,n_{adap}})$

- $\{\mathcal{H}_i^{j+1}\}_{i=1,n_{adap}} = \texttt{GenerateAdaptedMeshes}(\{\mathcal{H}_i^j\}_{i=1,n_{adap}}, \{\mathcal{M}_i^j\}_{i=1,n_{adap}})$

End for

Let us describe this algorithm also sketched in Figure 7.6. It consists of splitting the time interval $[0, T]$ into the n_{adap} mesh-adaptation time sub-intervals: $\{[t_{i-1}, t_i]\}_{i=1,\ldots,n_{adap}}$ with $t_0 = 0$ and $t_{n_{adap}} = T$. On each sub-interval, a different mesh is used. A time-forward computation of the state solution is first performed with an out-of-core storage of all checkpoints, which are taken to be $\{\mathcal{W}_h(t_i)\}_{i=1,\ldots,n_{adap}}$. Between each sub-interval, the solution is interpolated on the new mesh using the conservative interpolation of Alauzet and Mehrenberger (2010). Then, starting from the last sub-interval and proceeding until the first one, we recompute and store in memory all solution states of the sub-interval from the previously stored checkpoint in order to evaluate time-backward the adjoint state throughout the sub-interval. At the same time, we evaluate the Hessian metrics $\mathbf{H}_{\max}^i$ required to generate new adapted meshes for each sub-interval. To this end, a number

n_k of Hessian metric samples are computed on each sub-interval and intersected in the sense of the *metric intersection* defined in section 3.6.1 to obtain $\mathbf{H}^i_{\max}$. At the end of the computation, the global space–time mesh complexity is evaluated, producing weights for the GO metric fields for each sub-interval. Finally, all new adapted meshes are generated according to the prescribed metrics. The time-forward, time-backward and mesh update steps are repeated into the $j = 1, .., n_{ptfx}$ global fixed point loop. Convergence of the fixed point is obtained in typically five global iterations.

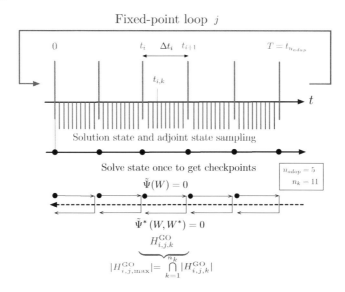

Figure 7.6. *Global transient fixed-point algorithm for unsteady goal-oriented anisotropic mesh adaptation. For a color version of this figure, see www.iste.co.uk/dervieux/meshadaptation2*

This mesh adaptation loop has been fully parallelized. The solution transfer, the solver and the Hessians computation nave been parallelized using a p-thread paradigm at the element loop level (Alauzet and Loseille 2009b). With regard to the computation of the metrics and the generation of the adapted meshes, we observe that they can be achieved in a decoupled manner once the complete simulation has been performed, leading to an easy parallelization of these stages. Indeed, if n_{adap} processors are available, each metric and mesh can be done on one processor.

7.6.3. *Computing the GO metric*

The optimal GO metric is a function of the adjoint state values, the adjoint state spatial gradients, the state time derivative, Hessian and the Euler fluxes Hessians. In practice, these continuous states are approximated by the discrete states and *derivative recovery* is applied to get gradients and Hessians. The discrete adjoint state W_h^* is taken to represent the adjoint state W^*. The gradient of the adjoint state ∇W^* is replaced by $\nabla_R W_h^*$ and the Hessian of each component of the flux vector $H(\mathcal{F}_i(W))$ is obtained from $H_R(\mathcal{F}_i(W_h))$. ∇_R (respectively, H_R) stands for the operator that *recovers* numerically the first (respectively, second) order derivatives of an initial piecewise linear solution field. In this chapter, the recovery method is based on the $\mathbf{L}^2$-projection formula, as explained in section 5.4.

7.7. Numerical experiments

The adaptation algorithms described in this chapter have implemented the CFD code Wolf. We use Yams (Frey 2001) for the adaptation in 2D and Feflo.a (Loseille and Löhner 2010) in 3D.

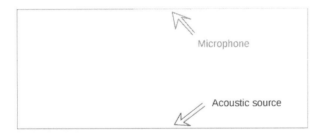

Figure 7.7. *Propagation in a box: sketch of geometry. An acoustic source produces sound at the center of the bottom. The microphone integrates the pressure variation on a segment of the top wall. For a color version of this figure, see www.iste.co.uk/dervieux/meshadaptation2*

7.7.1. *2D Acoustic wave propagation*

A typical example of pressure deviation propagating over long distances is linear acoustics. Linear acoustic waves usually refer either to a transient wave of bounded duration or to a periodic vibration. An important context in the study of these different kinds of waves is when we are interested only in the effect, during a limited time-interval, of the wave on a microphone occupying a very small part of the region affected by the pressure perturbation. Further simplifying, we can be interested in a single scalar measure of this effect, for instance the total energy E_{tot} received by the sensor during a given time interval. If the pressure perturbation is emitted at a very long distance in an open and complex spatial domain, the numerical simulation of this phenomenon can be extremely computer intensive, if not impossible. In Belme et al. (2012), the authors consider an acoustic source **s** ($A = 0.01, B = 256, C = 2.5$ and $f = 2$):

$$\mathbf{s}(\mathbf{x},t) = (0,0,0,r(\mathbf{x},t)) \text{ with } r(\mathbf{x},t)) = -Ae^{-Bln(2)[x^2+y^2]}Ccos(2\pi ft).$$

[7.42]

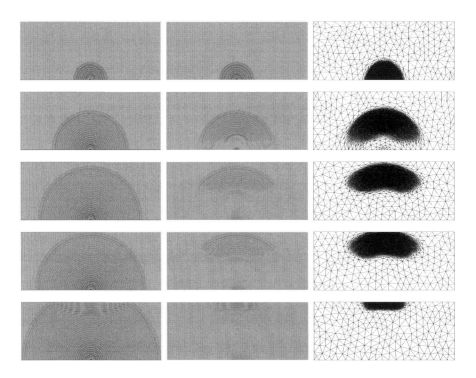

Figure 7.8. *Propagation of acoustic waves: density field evolving in time on a uniform mesh (left) and on adapted meshes (middle and right). For a color version of this figure, see www.iste.co.uk/dervieux/meshadaptation2*

We analyze the sound signal emitted by this source on a microphone M located at the center-top of the domain. The GO mesh adaptation considers the cost function

$$j(W) = \int_0^T \int_M \frac{1}{2}(p(\mathbf{x},t) - p_0)^2 \, \mathrm{d}M\mathrm{d}t.hv.$$

The time interval of the simulation is split into 40 sub-intervals (thus 40 checkpoints are used) and six fixed-point iterations are applied in order to converge the mesh adaptation problem. As expected, mesh adaptation reduces as much as possible mesh fineness in parts of the computational domain where accuracy loss does not influence the quality of sound prediction on the microphone. This is illustrated in Figure 7.8. The numerical convergence order is a central measure for evaluating the quality of adaptation. The scalar which we observe is the integrand $k(t)$ of the functional on the micro M, $k(t) = \int_M (p - p_0) \mathrm{d}M$. More precisely, we measure the amplitude of the first wave attaining M. With a sequence of uniform meshes with 60,000, 80,000 and 117,000 vertices, we obtain a poor convergence order of 0.6, while with a sequence of adapted meshes with 1,200, 24,000 and 64,000 vertices in the average, the spatial convergence order is 1.98.

7.7.2. *3D blast wave propagation*

Finally, the last example, also taken from Belme et al. (2012), considers a three-dimensional blast wave propagation in a complex geometry representing a city square surrounded by buildings. On the floor of the square we have a small half-sphere with high pressure and high density while the rest of atmosphere is standard. Cost function j is again the quadratic deviation from ambient pressure on target surface Γ, which is composed of one particular building (Figure 7.9):

$$j(W) = \int_0^T \int_\Gamma \frac{1}{2}(p(\mathbf{x}, t) - p_{air})^2 \mathrm{d}\Gamma \mathrm{d}t.$$

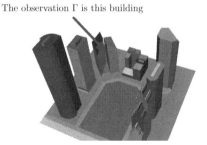

Figure 7.9. *3D City test case geometry and location of target surface Γ (surface of the building indicated by the arrow). For a color version of this figure, see www.iste.co.uk/dervieux/meshadaptation2*

The simulation time frame is split into 40 time sub-intervals, that is, 40 different adapted meshes are used to perform the simulation. Six fixed-point iterations are performed to converge the mesh adaptation problem. For each sub-interval, 16 samples of the solution and adjoint states are considered to build the GO metric. The space–time complexity is prescribed at 1.2 million.

The resulting adjoint-based anisotropic adapted meshes (surface and volume) for both simulations at sub-interval 10, 15 and 20 are shown in Figure 7.10. Despite the complexity and the unpredictable behavior of the physical phenomena with a large number of shock waves interactions with the geometry, the GO fixed point mesh adaptation algorithm is automatically able to capture those shock waves, which impact the targeted buildings and to set appropriate weights for the refinement of these physical phenomena. Other waves are neglected thus leading to a drastic reduction of the mesh size. To illustrate this point, we provide meshe sizes for sub-intervals 1, 5, 10 and 20 in Table 7.1, where nv, nt and nf are the number of vertices, tetrahedra and triangles, respectively, and h the mesh size. If, for sub-interval 1, 1 million vertices is needed, only 40,000 vertices are used for sub-interval 20. On average, almost 200,000 vertices are required with a maximal accuracy between 1 and 5 cm. These numbers have to be compared with 33,000,000, the number of vertices for an uniform mesh with accuracy of the order of cm. With regard to the amount of anisotropy achieved for these simulations, mesh anisotropy can be quantified by two different indicators: the anisotropic ratios and quotients, according to definition 3.19. The anisotropic ratio stands for the maximum elongation of a tetrahedron by comparing two principal directions. For both simulations, an average anisotropic ratio between 7 and 16 is achieved. The anisotropic quotient represents the overall anisotropic ratio of a tetrahedron taking into account all the possible directions, we get a mean anisotropic quotient between 40 and 200. This quotient can be considered as a measure of the overall gain in three dimensions of *an anisotropic adapted mesh* as compared to *an isotropic adapted one*.

Iteration	nv	nt	nf	min h	Ratio avg. (max)	Quotient avg. (max)
1	1,058,084	6,177,061	79,862	1. cm	9 (62)	63 (2,242)
10	249,620	1,427,956	34,318	1.2 cm	16 (94)	197 (5,136)
15	72,432	392,970	24,678	2.1 cm	12 (76)	119 (3,789)
20	40,297	205,855	21,980	4.8 cm	7 (64)	39 (3,547)
Avg.	188,357	1,074,777	28,811			

Table 7.1. *Mesh characteristics of the 3D blast wave calculation. For anisotropic ratio and quotient of an element, see definition 3.19 in Volume 1*

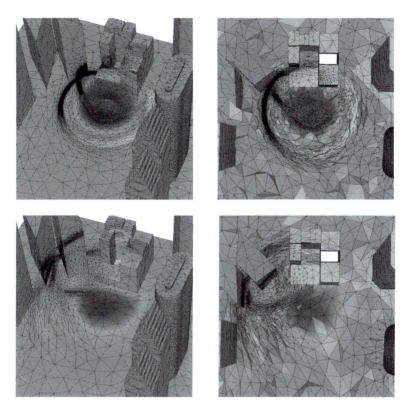

Figure 7.10. *3D Blast wave propagation: adjoint-based adapted surface (left) and volume (right) meshes at sub-intervals 10 and 20 and corresponding solution density at a-dimensioned times 5 and 10. For a color version of this figure, see www.iste.co.uk/dervieux/meshadaptation2*

7.8. Conclusion

This chapter introduces the GTFP mesh adaptation algorithm, which involves the specification of the set of spatial meshes used in an unsteady simulation as the optimum of a GO error analysis. This method specifies both mesh density and mesh anisotropy by variational calculus. Accounting for unsteadiness is applied in a time-implicit mesh-solution coupling, which needs a nonlinear iteration, the fixed point. In contrast to the Hessian-based fixed-point mentioned in Chapter 2 of this volume that iterates on each sub-interval, the new GTFP covers the whole time interval, including forward steps for evaluating the state and backward ones for the adjoint. This algorithm successfully applies to 2D and 3D blast wave Euler test cases and to the calculation of a 2D acoustic wave. Results demonstrate the favorable behavior expected from an adjoint-based method, which gives an automatic selection of the mesh regions necessary for the target output.

Several important issues for fully space–time computation have been addressed. Among them, the strategies for choosing the splitting in time sub-intervals and the accurate integration of time errors in the mesh adaptation process have been proposed, together with a more general formulation of the mesh optimization problem.

Time discretization error is not considered in this study. Solving this question is not so important for the type of calculations that are shown in this chapter, but can be of paramount impact in many other cases, in particular when implicit time advancing is considered.

7.9. Notes

More computational examples are given in Belme et al. (2012).

Link with further work. As mentioned in Chapter 4 of this volume, the proposed error analysis relies on the interpolation errors committed on the 15 Euler fluxes. In Chapter 5 of this volume, we introduce a linearized version in which it is enough to consider the interpolation errors on conserved unknowns. Although the linearized version has been tested only on steady viscous flow, it is probable that its accuracy in the unsteady viscous context can be better than the accuracy of the analysis presented in this chapter.

8
Third-Order Unsteady Adaptation

The quest for efficiency leads to combine mesh adaptation with higher order accuracy. This chapter presents a first attempt toward such a combination. A third-order accurate Central Essentially Non-Oscillatory (CENO) approximation is chosen for the demonstration. An *a priori error analysis* is developed. Then, using a least square projection of the error to a metric-based error reduces the *mesh adaptation goal-oriented problem* to the optimization of a metric. The numerical illustration concerns nonlinear sound propagation.

8.1. Introduction

Second-order mesh-adaptative approaches bring a good level of safety in the obtention of converging approximate solutions. But these solutions finally obtained are in best case second-order converging to the exact solution. On the other side, high-order solutions are much more efficient when they converge at high order, but often, convergence at high order is not obtained, either because mesh is not sufficient, or because the solution is singular or has stiff variations. Therefore, the Graal for getting accuracy and safety is to combine mesh adaptation and higher order approximations. In order to present a method combining both, we shall choose a higher order approximation. The recent investigations with unstructured meshes concern mainly discontinuous Galerkin (DG) approximations. DG is quite different (data structures, etc.) from the approximations considered in this book. Also, we are interested by singularities, for example, shocks, and, depending on the location of the shock in the mesh, the capture of the discontinuity by DG can be of good or less good quality. These two problems are avoided by a method like ENO. In this chapter, we consider mesh adaptation for a third-order accurate CENO approximation based on a quadratic polynomial reconstruction (then CENO2), which is used to approximate the unsteady Euler equations. This scheme is inspired by the CENO proposition of Ivan and Groth (2014; see also Yan and Ollivier-Gooch 2017). We transpose it beforehand into a vertex formulation. Because the third-order upwind scheme is rather dissipative, a modification is proposed in order to improve the approximation properties of the scheme according to the above criteria, focusing on its advective performances. As for the original unstructured CENO scheme (Charest et al. 2015), the new scheme is third-order accurate on irregular unstructured meshes.

Section 8.2 examines how to evaluate reconstruction errors. Section 8.3 presents shortly the numerical scheme, which has been designed for our investigation. Section 8.4 develops an error estimate based on the PDE, in order to build a goal-oriented mesh adaptation algorithm. The resulting a priori error analysis is a kind of dual of the a posteriori analysis of Barth and Larson (2002). Section 8.5 considers the extension of a metric-based adaptation method to take into account the cubic reconstruction error. Cao (2007b) proposes a first approximation of this error with stretching directions. In the present chapter, we propose to replace the application of the third derivative tensor to a mesh size vector by the power $3/2$ of the application of a pseudo-Hessian second-order tensor to this mesh size. The optimal metric is defined in section 8.6. The practical approach for an unsteady model is considered in section 8.7. For solving the resulting mesh optimality system, we discretize it and apply the global unsteady fixed point Algorithm 8.1. In section 8.8, the unsteady method is applied to an acoustic propagation benchmark and compared with previous approaches.

8.2. Higher order interpolation and reconstruction

Most high-order approximation schemes like DG (Bassi and Rebay 1997; Cockburn et al. 2000; Cockburn 2001; Shu and Cockburn 2001), ENO (Engquist et al. 1986; Barth and Frederickson 1990; Lafon and Abgrall Novembre 1993; Groth and Ivan 2011) or distributive schemes (Abgrall 2006) use kth-order interpolation or reconstruction and are k-exact in the sense that polynomial solutions of degree k are approximated without any error by such schemes. Interpolation and reconstruction are two approximation mappings, the errors of which need to be analyzed. Most analyses are inspired by the Bramble–Hilbert principle, stating that an approximation that is exact for kth-order polynomial is a $(k+1)$th-order accurate approximation. Demonstrations can be found in the fundamental paper (Ciarlet and Raviart 1972). Later, when considering reconstruction-based schemes (see Harten and Chakravarthy 1991), the authors referred to the Taylor series. A re-visitation in Abgrall (1992) establishes the link with Ciarlet and Raviart (1972). Interpolation errors are used for building adaptation criteria in Huang (2005). Several metrics are derived from the Hessians of each partial derivative. Then the metrics are intersected. A similar idea is presented in Hecht (2008). Intersections of metrics do not easily produce optimal meshes. They often result in a too severe anisotropy loss. A true asymptotic extension is proposed in Cao (2005, 2008). We also refer to Mirebeau (2010) for similar ideas. A singular Sylvester decomposition is applied in Mbinky (2013).

Let us focus in this section on the estimation of the *reconstruction error*. Given a function with sufficiently smooth u defined on a bounded domain Ω limited by a continuous boundary, given a tesselation of Ω into cells C_i of centroids $\mathbf{c}_i$, and the array $\bar{u} = \{\bar{u}_i\}$ of means of u on cells C_i, we are interested by polynomials $R_i(\mathbf{x}, \bar{u})$ of degree k built on any cell i and of same mean on cell i as u:

$$R_i(\mathbf{x}, \bar{u}) = \sum_{m=0}^{k} \frac{1}{m!} \sum_{|\ell|=m} (\mathbf{x} - \mathbf{c}_i)^\ell D_\ell, \quad \forall \mathbf{x} \in \Omega,$$

$$\times \int_{C_j} R_i(\mathbf{x}, \bar{u}) dV = \bar{u}_j, \quad \forall j \in J(i),$$

where $J(i)$ a set of cells close to cell C_i and ℓ holds for the usual multi-index notation, in 2D:

$$\ell = (\alpha_1, \alpha_2), ; \ (\mathbf{x} - \mathbf{c}_i)^\ell = (x - c_i^x)^{\alpha_1}(y - c_i^y)^{\alpha_2}; \ D_\ell = \frac{\partial^{\alpha_1}}{\partial x^{\alpha_1}} \frac{\partial^{\alpha_2}}{\partial y^{\alpha_2}}.$$

According to Harten and Chakravarthy (1991), for a sufficiently large neighborhood $J(i)$ of cells around i, we have

$$D_\ell = \frac{\partial^\ell u}{\partial x_\ell}(\mathbf{c}_i) + O(h^{k+1-|\ell|}) \quad \text{and} \quad R_i(\mathbf{x}, \bar{u}) = u(\mathbf{x}) + O(h^{k+1}),$$

where diameters of cells are less than h. Note that if we define the operator

$$\Pi : u \mapsto \Pi u, \quad \forall i, \ \forall \mathbf{x} \in C_i, \ \Pi u(\mathbf{x}) = R_i(\mathbf{x}, \bar{u}).$$

then, as far as the $J(i)$'s are sufficiently large, Π is k-exact (mapping a kth-order polynomial in itself). In Abgrall (1992), the authors use a result from Ciarlet and Raviart (1972) to give a more accurate estimate in the Sobolev space $W^{m,p}(\Omega)$ equipped of the following norm and semi-norm: $||.||_{m,p,\Omega}$,

$$||u||_{m,p,\Omega} = \left(\sum_{|\ell|=0}^{|\ell|=m} ||D^\ell u||_{p,\Omega}^p \right)^{\frac{1}{p}}, \quad |u|_{m,p,\Omega} = \left(\sum_{|\ell|=m} ||D^\ell u||_{p,\Omega}^p \right)^{\frac{1}{p}}.$$

It is written as

$$||u - \Pi u||_{m,p,\Omega} \leq C |u|_{k+1,p,\Omega} \frac{h^{k+1}}{\rho^m},$$

for a certain constant C and where ρ is related to the shape of cells and m is any integer such that $0 \leq m \leq k+1$. This shows that the reconstruction error is, when $k = 2$, effectively expressed in terms of the third derivative of u.

A last remark is that in the case of a mean-square-based reconstruction, if the number of cells of the support is exactly the number of unknown coefficients, then the minimum of the least square functional is zero that shows (for a smooth function) that the reconstruction is equal to the initial function in one point of each neighboring cell; in other words, the reconstruction is an interpolation, the error of which is given by the Taylor expansion. In the general case, we do not have a precise estimate, and we choose to get inspired by the Taylor expansion and heuristically write our reconstruction error estimate as follows:

$$\forall \mathbf{x}, \ \|u - \Pi u(\mathbf{x})\| \preceq \frac{1}{3!} \sup_{\delta \mathbf{x} \in \mathcal{B}(\mathbf{x})} |D^3 u(\delta \mathbf{x})^3|, \qquad [8.1]$$

where $\preceq$ holds for an inequality that holds for mesh size sufficiently small and where the $\delta \mathbf{x}$ describes the unit ball $\mathcal{B}(\mathbf{x})$ of local mesh sizes around $\mathbf{x}$, as defined in definition 3.18.

8.3. CENO approximation for the 2D Euler equations

8.3.1. *Model*

The unsteady 2D Euler equations in a geometrical bounded domain $\Omega \subset \mathbb{R}^2$ of boundary Γ can be written as

$$\text{Find } u \in \mathcal{V} \text{ such that } \int_0^T \int_\Omega \left(v \frac{\partial u}{\partial t} + v \nabla \cdot \mathcal{F}(u)\right) d\Omega \, dt$$

$$= \int_0^T \int_\Gamma v \mathcal{F}_\Gamma(u) \, d\Gamma \, dt, \quad \forall \, v \in \mathcal{V}. \qquad [8.2]$$

Here, $\mathcal{V} = L^2\big(0,T; H^1(\Omega)\big) \cap H^1\big(0,T; L^2(\Omega)\big)$ and $u = (u_1, u_2, u_3, u_4)$ hold for the conserved unknowns (density, moments components, energy) and $\nabla \cdot \mathcal{F} = (\nabla \cdot \overline{\mathcal{F}_1}, \nabla \cdot \overline{\mathcal{F}_2}, \nabla \cdot \overline{\mathcal{F}_3}, \nabla \cdot \overline{\mathcal{F}_4})$ with

$$\mathcal{F} = (\overline{\mathcal{F}_1^x}, \overline{\mathcal{F}_1^y}, \overline{\mathcal{F}_2^x}, \overline{\mathcal{F}_2^y}, \overline{\mathcal{F}_3^x}, \overline{\mathcal{F}_3^y}, \overline{\mathcal{F}_4^x}, \overline{\mathcal{F}_4^y}) = (\mathcal{F}_1, ..., \mathcal{F}_8) \qquad [8.3]$$

for the usual four Euler fluxes of mass, moments and energy. At right-hand side, we have an integral of the various boundary fluxes $\mathcal{F}_\Gamma$ for various boundary conditions, which we do not need to detail here. Defining

$$B(u,v) = \int_0^T \int_\Omega \left(v\frac{\partial u}{\partial t} + v\nabla \cdot \mathcal{F}(u)\right) d\Omega \, dt - \int_0^T \int_\Gamma v\mathcal{F}_\Gamma(u) \, d\Gamma \, dt,$$

the variational formulation is written as follows:

Find $u \in \mathcal{V}$ such that $B(u,v) = 0, \quad \forall \, v \in \mathcal{V}$. [8.4]

8.3.2. *CENO formulation*

We choose a reconstruction-based finite-volume method, being inspired by the unlimited version of the reconstruction technique of Barth (1993) and the CENO methods. Concerning the location of the nodes with respect to the mesh elements, we prefer to minimize the number of unknowns with respect to a given mesh; therefore we keep the *vertex-centered* location already used in this book for second-order anisotropic (Hessian-based or goal-oriented) mesh adaptation (Chapter 5 of Volume 1 and Chapters 2, 4 and 7 of Volume 2). For a more detailed description of the CENO approach presented here, see Ouvrard et al. (2009). Its main features are as follows: (a) vertex centered, (b) dual median cells around the vertex, (c) a single mean square quadratic reconstruction for each dual cell, (d) Roe approximate Riemann solver for stabilization and (e) explicit multi-stage time-stepping. The computational domain is divided in triangles and in a dual tesselation in cells, each cell C_i being built around a vertex i, with limits following sections of triangle medians (Figure 8.1). We define the discrete space $\mathcal{V}_0$ of functions of $\mathcal{V}$, which are constant on any dual cell C_i. Let us define a *discrete reconstruction operator* R_2^0. The operator R_2^0 reconstructs a function of $\mathcal{V}_0$ in each cell C_i under the form of a second-order polynomial: $R_2^0 u_0|_{C_i} = \mathcal{P}_i^2(\mathbf{x})$. Given the means $(\overline{u_{0,i}}, i = 1, ...)$ of u_0 on cells C_i, $\mathcal{P}_i^2(\mathbf{x})$ is defined by the $c_{i,\alpha}, |\alpha| \leq k$ such that

$$\mathcal{P}_i^2(\mathbf{x}) = \overline{u_{0,i}} + \sum_{|\alpha|\leq k} c_{i,\alpha}[(\mathbf{x} - \mathbf{c}_i)^\alpha - \overline{(\mathbf{x} - \mathbf{c}_i)^\alpha}]$$

$$\overline{\mathcal{P}_{i,i}} = \overline{u_{0,i}}; \ (c_{i,\alpha}, |\alpha| \leq k) = Arg \min \sum_{j \in N(i)} (\overline{\mathcal{P}_{i,j}} - \overline{u_{0,j}})^2,$$

where $\overline{\mathcal{P}_{i,j}}$ stands for the mean of $\mathcal{P}_i^2(\mathbf{x})$ on cell j, and the set of neighboring cells is taken sufficiently large for an accurate quadratic reconstruction (see Figure 8.1).

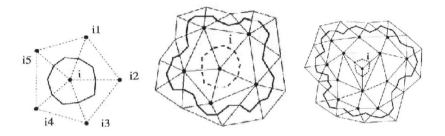

Figure 8.1. *Dual cell and two reconstruction molecules*

For the Euler model [8.4], the semi-discretized CENO scheme is written as follows:

Find $u_0 \in \mathcal{C}^1([0,T]; \mathcal{V}_0)$ such that $B(R_2^0 u_0, v_0) = 0 \vee v_0 \in \mathcal{C}^1([0,T]; \mathcal{V}_0)$ with

$$B_h(R_2^0 u_0, v_0) = \int_{\Omega \times [0,T]} v_0 \frac{\partial R_2^0 u_0}{\partial t} \, d\Omega + \int_{\Omega \times [0,T]} v_0 \nabla_h \cdot \mathcal{F}(R_2^0 u_0) \, d\Omega dt$$

$$- \int_{\Gamma \times [0,T]} v_0 \mathcal{F}_\Gamma(R_2^0 u_0) \, d\Gamma \, dt. \qquad [8.5]$$

In [8.5], the term $\nabla_h \cdot \mathcal{F}(R_2^0 u_0)$ needs to be defined, since $\mathcal{F}(R_2^0 u_0)$ is discontinuous at cells interfaces. Taking v_0 as a characteristic function of cell C_i, we get the following finite volume formulation:

$$\forall C_i, \int_{C_i} \frac{\partial R_2^0 u_0}{\partial t} \, d\Omega + \int_{C_i} \nabla_h \cdot \mathcal{F}(R_2^0 u_0) \, d\Omega \int_{\partial C_i \cap \Gamma} \mathcal{F}_\Gamma(R_2^0 u_0) \, d\Gamma = 0 \quad \forall t$$

or, (using $\overline{\mathcal{P}_{i,i}} = \overline{u_{0,i}}$),

$$\forall C_i, \frac{\partial}{\partial t} \int_{C_i} u_0 \, d\Omega + \int_{\partial C_i} \mathcal{F}(R_2^0 u_0) \cdot \mathbf{n} \, d\Gamma - \int_{\partial C_i \cap \Gamma} \mathcal{F}_\Gamma(R_2^0 u_0) \, d\Gamma = 0. \qquad [8.6]$$

Neither the discrete divergence $\nabla_h \cdot$ nor the CENO approximation is defined by the definition of the reconstruction. Indeed, the reconstruction performed in each cell produces a global field, which is generally discontinuous at cell interfaces $\partial C_i \cap \partial C_j$

(see Figure 8.2). In order to fix an integration value at the interface, we can consider an arithmetic mean of the fluxes values for the two reconstruction values:

$$\mathcal{F}(R_2^0 u_0)^{quadrature}|_{\partial C_i \cap \partial C_j} \cdot \mathbf{n} = \frac{1}{2}\left(\mathcal{F}(R_2^0 u_0)|_{\partial C_i} + \mathcal{F}(R_2^0 u_0)|_{\partial C_j}\right) \cdot \mathbf{n}, \qquad [8.7]$$

where $(R_2^0 u_0)|_{\partial C_i}$ holds for the value at cell boundary of the reconstructed $R_2^0 u_0|_{C_i}$ on cell C_i. The above mean is applied on the four Gauss integration points ($\mathbf{g}_\alpha, \alpha = 1, 4$) (two per interface segment, see Figure 8.2) necessary for an exact integration of quadratic polynomials. Then the accurate definition of B_h is as follows:

$$B_h(R_2^0 u_0, v_0) = \int_0^T \left\{ \sum_i \left[\int_{C_i} v_0 \frac{\partial u_0}{\partial t} d\Omega \right. \right.$$
$$\left. + \int_{\partial C_i} v_0 \sum_{\alpha=0}^{\alpha=4} \frac{\mathcal{F}(R_2^0 u_0)|_{\partial C_i}(\mathbf{g}_\alpha) + \mathcal{F}(R_2^0 u_0)|_{\partial C_j}(\mathbf{g}_\alpha)}{8} d\sigma \right] \right\} dt$$
$$+ \int_0^T \int_{\partial \Gamma} v_0 d\Gamma dt . \qquad [8.8]$$

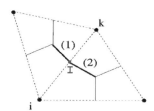

Figure 8.2. *Sketch of the interface $\partial C_i \cap \partial C_k$ between cell C_i and cell C_k. It is made of two segments (1) and (2) joining mid-edge I and triangles centroid. Flux integration on (1) (respectively, (2)) will rely on two Gauss integration points. For a color version of this figure, see www.iste.co.uk/dervieux/meshadaptation2*

The restriction of B_h to smooth function is B and for the continuous solution u satisfies:

$$B_h(u, v_0) = B(u, v_0) = 0 \quad \forall \, v_0 \in \mathcal{C}^1([0, T]; \mathcal{V}_0). \qquad [8.9]$$

It remains to define a time discretization for [8.6]. We apply the standard explicit Runge-Kutta (RK4) time advancing. With the above central-differenced [8.7] spatial quadrature, this formulation produces a central-differenced numerical approximation, which is third-order accurate. However, in a nonlinear setting or on irregular unstructured meshes, it cannot be used due to a lack of stability.

8.3.3. *Vertex-centered low dissipation CENO2*

Scheme [8.6] is usually combined with an approximate Riemann solver instead of the formulation [8.7] proposed in the previous section. This latter option produces the usual upwind-CENO scheme, which enjoys a rather good nonlinear stability, but shows in applications a relatively large dissipation error. Indeed, its 1D analysis put in evidence a dominant dissipation error term of weight Δx^3 and similar to the fourth spatial derivative of the unknown solution, as it is the case of usual second-order approximations. In contrast, using a four-degree reconstruction would produce an approximation scheme involving solely a dissipation by sixth spatial derivative, in Δx^5, much smaller. In the rest of this chapter, we shall use a low-dissipation version introduced in Carabias et al. (2011). The CENO with central differencing on integration points for fluxes is stabilized by the use on each edge ij of approximates polynomial 1D interpolations in the direction of ij. We refer to Carabias et al. (2011) for details. The reader just needs to know that the accuracy is impressively improved as sketched in Figure 8.3, which compares the advection of a Gaussian profile for the upwind CENO and the low-dissipation one. Both schemes are third-order accurate and the error analysis can be performed for both schemes without considering the low-dissipation modification.

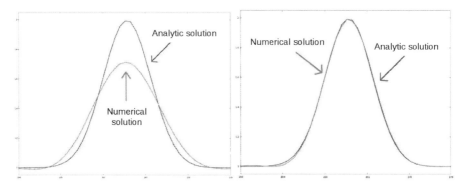

Figure 8.3. *Improvement of the CENO scheme for the advection of a 2D Gauss-shaped concentration through* 400 *space intervals. Left: Comparison of the upwind third-order accurate CENO solution with the analytic solution. Right: Comparison of the improved (third-order accurate) CENO scheme with the analytic solution. For a color version of this figure, see www.iste.co.uk/dervieux/meshadaptation2*

8.4. Error analysis

In the present work, we do not consider time discretization errors. For explicit high-order time-advancing subject to a Courant-type condition, some argument for discarding the time-error analysis of advective models can be found in the Annex of Chapter 2 of this volume. We then concentrate on spatial errors. Furthermore, we do not analyze the corrections terms introduced in previous section. At least on regular meshes, these corrections terms are terms of higher order. The proposed a priori analysis is in some manner the dual of the a posteriori analysis proposed by Barth and Larson (2002). Let $j(u) = (g, u)$ be the scalar output, which we want to accurately compute, where u is the solution of the continuous system [8.4]. We concentrate on the reduction, by mesh adaptation, of the following error:

$$\delta j = (g, R_2^0 \pi_0 u - R_2^0 u_0),$$

where g is function of $L^2(\Omega)$ and u_0 the discrete solution of [8.6]. The projection π_0 is defined by

$$\pi_0 : v \mapsto \pi_0 v, \quad \pi_0 v|_{C_i} = \int_{C_i} v dx \quad \forall\, C_i, \text{dual cell}.$$

The adjoint state $u_0^* \in \mathcal{C}^1([0,T]; \mathcal{V}_0)$ is the solution of B_h (defined in [8.5]):

$$\frac{\partial B_h}{\partial u}(R_2^0 u_0)(R_2^0 v_0, u_0^*) = (g, R_2^0 v_0), \ \forall\, v_0 \in \mathcal{V}_0. \qquad [8.10]$$

Then we can write, successively,

$$\begin{aligned}(g, R_2^0 \pi_0 u - R_2^0 u_0) &= \frac{\partial B_h}{\partial u}(R_2^0 u_0)(R_2^0 \pi_0 u - R_2^0 u_0, u_0^*) \text{ (adjoint eq.)[8.10]} \\ &\approx B_h(R_2^0 \pi_0 u, u_0^*) - B_h(R_2^0 u_0, u_0^*)\end{aligned}$$

and then

$$\begin{aligned}(g, R_2^0 \pi_0 u - R_2^0 u_0) &\approx B_h(R_2^0 \pi_0 u, u_0^*) \text{ (discrete state eq.)[8.5][8.7]} \\ &\approx B_h(R_2^0 \pi_0 u, u_0^*) - B_h(u, u_0^*) \text{ (continuous state eq.)[8.4] [8.9]} \\ &\approx \frac{\partial B_h}{\partial u}(u)(R_2^0 \pi_0 u - u, u_0^*).\end{aligned}$$

For the case of Euler equations, the previous error estimate is written as

$$\frac{\partial B_h}{\partial u}(u)(R_2^0\pi_0 u - u, u_0^*) =$$

$$\int_0^T \frac{\partial B_h^{time}}{\partial u}(u)(R_2^0\pi_0 u - u, u_0^*)dt + \int_0^T \frac{\partial B_h^{Euler}}{\partial u}(u)(R_2^0\pi_0 u - u, u_0^*)dt$$

with

$$\frac{\partial B_h^{time}}{\partial u}(u)(R_2^0\pi_0 u - u, u_0^*) = \sum_i \int_{C_i} u_0^*(R_2^0\pi_0 - Id)\frac{\partial u}{\partial t}dx$$

$$\frac{\partial B_h^{Euler}}{\partial u}(u)(R_2^0\pi_0 u - u, u_0^*) =$$

$$\sum_i \int_{C_i} u_0^* \nabla_h \cdot \mathcal{F}'(u)(R_2^0\pi_0 u \quad u)dx \quad - \int_{\partial C_i \cap \Gamma} u_0^* \mathcal{F}_1'(u)(R_2^0\pi_0 u - u)\, d\Gamma$$

with the sum applied for all dual cell C_i of the mesh. Since, as in Loseille et al. (2007), *we do not consider the adaptation of boundary mesh*, we discard the boundary terms:

$$\frac{\partial B_h^{Euler}}{\partial u}(u)(R_2^0\pi_0 u - u, u_0^*) \approx \sum_{i.} \sum_i \int_{C_i} u_0^* \nabla_h \cdot \mathcal{F}'(u)(R_2^0\pi_0 u - u)dx,$$

that is,

$$\frac{\partial B_h^{Euler}}{\partial u}(u)(R_2^0\pi_0 u - u, u_0^*) - \int_{C_i} u_0^* \frac{\partial u_0}{\partial t}d\Omega \approx \sum_{j \in \mathcal{V}(i)} \int_{\partial C_i \cap \partial C_j}$$

$$\times u_0^* \sum_{\alpha=0}^{\alpha=4} \frac{\mathcal{F}'(u)(R_2^0\pi_0 u - u)|_{\partial C_i}(\mathbf{g}_\alpha) + \mathcal{F}'(u)(R_2^0\pi_0 u - u)|_{\partial C_j}(\mathbf{g}_\alpha)}{8} \cdot \mathbf{n} d\sigma. \quad [8.11]$$

Let us examine the ingredients of the RHS integrand. First, u_0^* is constant by cell, with discontinuities at cell interface of amplitude of order of mesh size. The Jacobian $\mathcal{F}'(u)$ is smooth. The reconstruction error $R_2^0\pi_0 u - u$ is discontinuous at cell interfaces, but for a u smooth, the amplitude of this discontinuity is of order the third power of mesh size. Then

$$\mathcal{F}'(u)(R_2^0\pi_0 u - u)|_{\partial C_k}(\mathbf{g}_\alpha) \approx \mathcal{F}'(u)(R_2^0\pi_0 u - u)(\mathbf{G}_i) \quad k = i \text{ or } k = j \in \mathcal{V}(i),$$

where $\mathbf{G}_i$ is the centroid of cell C_i. Thus,

$$\frac{\partial B_h^{Euler}}{\partial u}(u)(R_2^0\pi_0 u - u, u_0^*) - \int_{C_i} u_0^* \frac{\partial u_0}{\partial t} d\Omega \approx$$
$$\mathcal{F}'(u)(R_2^0\pi_0 u - u)(\mathbf{G}_i) \cdot \sum_{j \in \mathcal{V}(i)} \int_{\partial C_i \cap \partial C_j} u_0^* \mathbf{n} d\sigma. \quad [8.12]$$

We recognize inside the RHS the approximation of a gradient of u_0^*:

$$\sum_{j \in \mathcal{V}(i)} \int_{\partial C_i \cap \partial C_j} u_0^* \mathbf{n} d\sigma \approx area(C_i) \boldsymbol{\nabla} u_0^*,$$

thus

$$\frac{\partial B_h^{Euler}}{\partial u}(u)(R_2^0\pi_0 u - u, u_0^*) - \int_{C_i} u_0^* \frac{\partial u_0}{\partial t} d\Omega \approx$$
$$area(C_i) \, \mathcal{F}'(u)(R_2^0\pi_0 u - u)(\mathbf{G}_i) \cdot \boldsymbol{\nabla} u_0^*.$$

As remarked in section 8.2, $R_2^0\pi_0 u - u$ can be replaced by a smooth function of the local third derivatives and local mesh size (factor $\frac{1}{3!}$ can be discarded in the rest of the analysis):

$$R_2^0\pi_0 u_q - u_q \preceq \sup_{\delta \mathbf{x}} |D^3 u(\mathbf{x})(\delta \mathbf{x})^3|, \ \forall \ q = 1, 4.$$

For each flux component ($r = 1, 8$),

$$\mathcal{F}_r'(R_2^0\pi_0 u - u) \preceq \sup_{\delta \mathbf{x}} \sum_q ||\mathcal{F}_{qr}'||(D^3 u_q(\mathbf{x})(\delta \mathbf{x})^3).$$

Then

$$|\delta j| \preceq \sum_q \int_\Omega K_q^{time}(u^*) \sup_{\delta \mathbf{x}} |D^3 \frac{\partial u}{\partial t}(\delta \mathbf{x})^3)| \, d\Omega$$
$$+ 2\sum_q \int_\Omega K_q^{Euler}(u, u^*) \sup_{\delta \mathbf{x}} |D^3 u(\delta \mathbf{x})^3)| \, d\Omega,$$

with

$$K_q^{time}(u^*) = |u^*| \quad ; \quad K_q^{Euler}(u,u^*) = \sum_r |(\mathcal{F}'_{rq}(u))^*| |\frac{\partial u_q^*}{\partial x_r}|,$$

with the notation $\mathcal{F}$ according to [8.3], and

$$\partial u_q^*/\partial x_r = \partial u_q^*/\partial x \quad \text{for} \quad q = 1, 3, 5, 7,$$
$$\partial u_q^*/\partial x_r = \partial u_q^*/\partial y \quad \text{for} \quad q = 2, 4, 6, 8.$$

It will be useful to write it as follows:

$$|\delta j| \preceq \int_\Omega \sup_{\delta \mathbf{x}} \mathbb{T}(|\delta \mathbf{x}|)^3 \, d\Omega,$$

where $\mathbb{T}$ is the trilinear tensor:

$$\mathbb{T}(|\delta \mathbf{x}|)^3 = \sum_q \left(K_q^{time}(u^*)|D^3 \frac{\partial u}{\partial t}|(|\delta \mathbf{x}|)^3 + K_q^{Euler}(u,u^*)|D^3 u|(|\delta \mathbf{x}|)^3 \right). \quad [8.13]$$

Not surprisingly, the error is a third-order tensor in terms of the $\delta \mathbf{x}$, measuring local mesh size. In order to apply a metric-based mesh adaptation, we shall convert the estimate in a pseudo-quadratic estimate.

8.5. Metric-based error estimate

Given a Riemannian metric $\mathcal{M}(\mathbf{x}) = d\,\mathcal{R}(\mathbf{x})\Lambda(\mathbf{x})\mathcal{R}^t(\mathbf{x})$ parameterizing the mesh (Chapter 3 of Volume 1), the P_1 interpolation error of a smooth (at least $\mathcal{C}^2$) function u_q satisfies the following:

$$|u_q(\mathbf{x}) - \pi_1^\mathcal{M} u_q(\mathbf{x})| \approx |\frac{\partial^2 u_q}{\partial \tau_q^2}|(\delta \tau_q)^2 + |\frac{\partial^2 u_q}{\partial n_q^2}|(\delta n_q)^2$$
$$= \delta \mathbf{x}_\mathcal{M} |H_{u_q}| \delta \mathbf{x}_\mathcal{M} = trace(\mathcal{M}^{-\frac{1}{2}}|H_{u_q}|\mathcal{M}^{-\frac{1}{2}}), \quad [8.14]$$

where H_{u_q} is the Hessian of u_q, and orthonormal directions $\tau_q = (\tau_x^q, \tau_y^q)$ and $n_q = (n_x^q, n_y^q)$ are eigenvectors of this Hessian and $\delta \mathbf{x}_\mathcal{M}$ a vector of $\mathbb{R}^2$ such that $\delta \mathbf{x}_\mathcal{M} \mathcal{M} \delta \mathbf{x}_\mathcal{M} = 1$. This formulation permits the research of an optimal metric minimizing the linear interpolation error (Chapter 4 of Volume 1). We have to adapt this to a higher order interpolation/reconstruction. In Mbinky et al. (2012), the authors propose a general statement for an interpolation of arbitrary degree generalizing [8.14]. We prefer here a simpler option. In order to address our third-order mesh adaptation problem in a similar manner to the Hessian-based approach, we propose instead using $\mathbb{T}$ to define on each vertex i a pseudo-Hessian $\tilde{H}_i$ in such a way that inside cell i we have (see [8.1])

$$\mathbb{T}(\mathbf{x}_i)(|\delta\mathbf{x}|)^3 \approx \left(|\tilde{H}_i|(\delta\mathbf{x})^2\right)^{\frac{3}{2}} \quad \forall \delta\mathbf{x} \in \mathcal{B}(\mathbf{x}).$$

In the continuous mesh model, the set of neighboring vertices j around a given vertex i is the surface of an ellipse centered on i and with small and large axis defined by the metric matrix $\mathcal{M}$. One way to adjust the pseudo-Hessian $\tilde{H}$ consistently with the discrete context is to minimize by least square projection the deviation between the quadratic term and the trilinear one. The ball $\mathcal{B}(\mathbf{x})$ is approximated by considering any edge ij around i in the current mesh:

$$\tilde{H}_i = Argmin \sum_{j=1}^{N(i)} \left(\tilde{H}_i(\vec{ij})^2 - (\mathbb{T}(\mathbf{x}_i)(\vec{ij})^3)^{2/3} \right)^2,$$

with $\vec{ij} = (x_j - x_i, y_j - y_i)$, $N(i)$ being the set of neighbors of i. Replacing then the trilinear estimate by the $3/2$ power of the quadratic term, we get

$$\sup_{\delta\mathbf{x}} \mathbb{T}(\mathbf{x}_i)(|\delta\mathbf{x}|)^3 \preceq \left(trace(\mathcal{M}^{-\frac{1}{2}}|\tilde{H}_i|\mathcal{M}^{-\frac{1}{2}})\right)^{\frac{3}{2}}.$$

We consider now a way to find a metric field which minimizes this error term.

8.6. Optimal metric

Because of the above transformation of the error estimate, we can consider the minimization of the following error functional derived from estimate [8.13]:

$$\mathcal{E}_0 = \sum_{q=1,4} \int_\Omega K_q(u, u^*) \left(trace(\mathcal{M}^{-\frac{1}{2}}|\tilde{H}_{u_q}|\mathcal{M}^{-\frac{1}{2}})\right)^{\frac{3}{2}} dxdy.$$

We then obtain

$$\mathcal{E}_0 \preceq \mathcal{E} = \int_\Omega \left(trace(\mathcal{M}^{-\frac{1}{2}}|S|\mathcal{M}^{-\frac{1}{2}})\right)^{\frac{3}{2}} dxdy$$
$$\text{with} \quad S = \sum_{q=1,4} K_q(u,u^*)^{\frac{2}{3}} |\tilde{H}_{u_q}|. \qquad [8.15]$$

Matrix $S(\mathbf{x})$ is a sum of symmetric positive definite matrices, thus $S(\mathbf{x}) = \mathcal{R}_S(\mathbf{x})\Lambda_S(\mathbf{x})\mathcal{R}_S^t(\mathbf{x})$. We now identify the optimal metric, $\mathcal{M}^{opt} = \mathcal{M}^{opt}(N)$, among those having a prescribed total complexity N, which minimizes the above error. We proceed similarly to the second-order metric analysis (section 4.4 of this volume). To do so, we re-write [8.15] as

$$\int_\Omega \left(trace(\mathcal{M}^{-\frac{1}{2}}|S|\mathcal{M}^{-\frac{1}{2}})\right)^{\frac{3}{2}} dxdy =$$
$$\int_\Omega \left(trace(d_\mathcal{M}^{-1}(\mathcal{R}_\mathcal{M}\Lambda_\mathcal{M}\mathcal{R}_\mathcal{M}^T)^{-\frac{1}{2}}|S|(\mathcal{R}_\mathcal{M}\Lambda_\mathcal{M}\mathcal{R}_\mathcal{M}^T)^{-\frac{1}{2}})\right)^{\frac{3}{2}} dxdy.$$

Mesh stretching direction. We first prescribe, at each point $\mathbf{x}$ of the computational domain Ω, the adapted metric eigenvectors, that is, the representation of the direction of stretching of mesh, $\mathcal{R}_{\mathcal{M}_{opt}}$ as aligned with the above error model, that is,

$$\mathcal{R}_{\mathcal{M}_{opt}} = \mathcal{R}_S.$$

Mesh stretching length. Let us minimize the error at each point $\mathbf{x}$ of the computational domain for a prescribed mesh density $d_\mathcal{M}$. We derive that the best ratio of eigenvalues for $\mathcal{M}$, that is, the representation of mesh stretching or anisotropy should uniformize the two components of the error. This means that the product

$$(\mathcal{R}_{\mathcal{M}_{opt}}\Lambda_{\mathcal{M}_{opt}}\mathcal{R}_{\mathcal{M}_{opt}}^T)^{-\frac{1}{2}}|S|(\mathcal{R}_{\mathcal{M}_{opt}}\Lambda_{\mathcal{M}_{opt}}\mathcal{R}_{\mathcal{M}_{opt}}^T)^{-\frac{1}{2}}$$

is made proportional to identity. It implies that

$$e_{opt} = \frac{(\lambda_{1_S})^{-\frac{1}{2}}}{(\lambda_{2_S})^{-\frac{1}{2}}} \quad ; \quad \Lambda_{\mathcal{M}_{opt}} = diag[e_{\mathcal{M}_{opt}}^{-1}, e_{\mathcal{M}_{opt}}],$$

in which we have enforced $det(\Lambda_{\mathcal{M}_{opt}}) = 1$. With these definitions of $\mathcal{R}_{\mathcal{M}_{opt}}$ and $\Lambda_{\mathcal{M}_{opt}}$, we get

$$(\mathcal{R}_{\mathcal{M}_{opt}}\Lambda_{\mathcal{M}_{opt}}\mathcal{R}_{\mathcal{M}_{opt}}^T)^{-\frac{1}{2}}|S|(\mathcal{R}_{\mathcal{M}_{opt}}\Lambda_{\mathcal{M}_{opt}}\mathcal{R}_{\mathcal{M}_{opt}}^T)^{-\frac{1}{2}} = \begin{pmatrix} \lambda_{1_S}^{\frac{1}{2}}\lambda_{2_S}^{\frac{1}{2}} & 0 \\ 0 & \lambda_{1_S}^{\frac{1}{2}}\lambda_{2_S}^{\frac{1}{2}} \end{pmatrix}.$$

Inside this restricted family of metrics, it remains to define the optimal metric density. Let us consider the set of metrics with a total number of vertices prescribed to N:

$$\int_\Omega d\,dxdy = N. \qquad [8.16]$$

We now have to minimize the L^1 norm of the error

$$\mathcal{E}(d) = \int_\Omega d^{-\frac{3}{2}}\Gamma(S)\,dxdy$$

$$\Gamma(S) = \left(trace\begin{pmatrix} \lambda_{1_S}^{\frac{1}{2}}\lambda_{2_S}^{\frac{1}{2}} & 0 \\ 0 & \lambda_{1_S}^{\frac{1}{2}}\lambda_{2_S}^{\frac{1}{2}} \end{pmatrix}\right)^{\frac{3}{2}} = (2\lambda_{1_S}^{\frac{1}{2}}\lambda_{2_S}^{\frac{1}{2}})^{\frac{3}{2}} \qquad [8.17]$$

with respect to d for a given number of nodes N. This means that

$$\mathcal{E}'(d)\cdot\delta d = 0 \ \forall\ \delta d\ \text{with}\ \int_\Omega \delta d\,dxdy = 0,$$

which implies that the derivative $\Gamma(S)d^{-\frac{5}{2}}$ of integrand in $\mathcal{E}$ is constant and produces the optimal density

$$d_{opt} = \frac{N}{C_{opt}}(\Gamma(S))^{\frac{2}{5}} = \frac{N}{C_{opt}}(2\lambda_{1_S}^{\frac{1}{2}}\lambda_{2_S}^{\frac{1}{2}})^{\frac{3}{5}}\ \text{with}\ C_{opt} = \int_\Omega(2\lambda_{1_S}^{\frac{1}{2}}\lambda_{2_S}^{\frac{1}{2}})^{\frac{3}{5}}\,dxdy.$$

This completes the definition of the optimal metric:

$$\mathcal{M}_{opt} = d_{opt}\mathcal{R}_{opt}^t\begin{pmatrix} e_{opt}^{-1} & 0 \\ 0 & e_{opt} \end{pmatrix}\mathcal{R}_{opt}.$$

8.7. From theory to practical application

The previous analysis sets the optimal metric problem as the solution of the following continuous optimality conditions:

$$u \in \mathcal{V}, \ \forall \varphi \in \mathcal{V}, \ B(u, \varphi) = 0 \qquad \text{"state system"}$$

$$u^* \in \mathcal{V}, \ \forall \psi \in \mathcal{V}, \ \left(\frac{\partial B}{\partial u}\right)(u, u^*)\psi = (g, \psi) \qquad \text{"adjoint system"}$$

$$\mathcal{M}(\mathbf{x}, t) = \mathcal{M}_{opt}(\mathbf{x}, t) \qquad \text{"control optimality"}.$$

This system cannot be solved analytically and will be finally discretized. Not surprisingly, we choose the discrete state system as the one introduced previously. The adjoint is derived from its linearization. The last equation is discretized by recovering third-order derivatives by using a fourth-order reconstruction (like it would be done for a CENO3 scheme). Then, for each time step, the nonlinear iteration on the three equations of the optimality system provides an optimal metric and a unit mesh. In order to avoid to remesh too frequently, we adapt the Global Transient Fixed Point algorithm introduced in Chapter 7 of this volume and sketched in Figure 7.6. We keep its important features. The time interval has three embedded discretizations: (1) time sub-intervals $]t_i, t_{i+1}[$ during which the mesh is frozen, (2) divisions $]t_{i,k}, t_{i,k+1}[$ of these sub-intervals for checkpointing state solution to be re-used backward in time for solving the adjoint system and (3) an even finer discretization in the time-steps $]t_{i,k} + m\delta t, t_{i,k} + (m+1)\delta t[$ used to advance in time. For each subinterval n_{adap}, each maximal metric $|\mathbf{H}_{\max}|_i^j$ is computed from the different variables and different time levels of the subinterval by applying the metric intersection defined in section 3.6.1 of Chapter 3 of Volume 1. We rewrite the adapted algorithm with the modification for third order in Algorithm 8.1. A crucial criterion for evaluating a mesh adaptation method is its a priori/theoretical convergence order in terms of used degrees of freedom. Similarly, the evaluation of a computation should also rely on numerical convergence order, which should be preferably close to the theoretical one. A detailed discussion of the convergence of the global transient fixed point is presented in section 2.9.2 of Volume 1. According to lemma 2.12, the *space-time theoretical order* (i.e. in terms of space and time degrees of freedom) is for an arbitrary flow (possibly with singularities) limited to 9/5 in 2D and 2 in 3D. In contrast, the *space theoretical order*, evaluated by considering solely spatial degrees of freedom maybe as high as the theoretical spatial truncation order. It can be measured by solely considering the mean number of vertices (for a fixed number of mesh-adaptation intervals).

Algorithm 8.1. Third-order global fixed-point

```
    - For j=1,nptfx  //--- Solve state once to get checkpoints
For i=1,nadap
```

- $\mathcal{S}_{0,i}^{j} = \texttt{ConservativeSolutionTransfer}(\mathcal{H}_{i-1}^{j}, \mathcal{S}_{i-1}^{j}, \mathcal{H}_{i}^{j})$
- $\mathcal{S}_{i}^{j} = \texttt{SolveStateForward}(\mathcal{S}_{0,i}^{j}, \mathcal{H}_{i}^{j})$

```
    End for   //--- Solve state forward and adjoint backward in time and
store samples
For i=nadap,1
```

- $(\mathcal{S}^{*})_{i}^{j} = \texttt{AdjointStateTransfer}(\mathcal{H}_{i+1}^{j}, (\mathcal{S}_{0}^{*})_{i+1}^{j}, \mathcal{H}_{i}^{j})$
- $\{\mathcal{S}_{i}^{j}(k), (\mathcal{S}^{*})_{i}^{j}(k)\}_{k=1,nk} = \texttt{SolveStateAndAdjointBackward}(\mathcal{S}_{0,i}^{j}, (\mathcal{S}^{*})_{i}^{j}, \mathcal{H}_{i}^{j})$
- $|\mathbf{T}_{\max}|_{i}^{j} = \texttt{ComputeGoalOriented3rdOrdertensor}(\mathcal{H}_{i}^{j}, \{\mathcal{S}_{i}^{j}(k), (\mathcal{S}^{*})_{i}^{j}(k)\}_{k=1,nk})$
- $|\mathbf{H}_{\max}|_{i}^{j} = \texttt{ComputeGoalOrientedHessianMetric}(\mathcal{H}_{i}^{j}, \{\mathcal{S}_{i}^{j}(k), (\mathcal{S}^{*})_{i}^{j}(k)\}_{k=1,nk})$

```
    End for
```
- $\mathcal{C}^{j} = \texttt{ComputeSpaceTimeComplexity}(\{|\mathbf{H}_{\max}|_{i}^{j}\}_{i=1,nadap})$
- $\{\mathcal{M}_{i}^{j-1}\}_{i=1,nadap} = \texttt{ComputeUnsteadyGoalOrientedMetrics}(\mathcal{C}^{j-1}, \{|\mathbf{H}_{\max}|_{i}^{j-1}\}_{i=1,nadap})$
- $\{\mathcal{H}_{i}^{j}\}_{i=1,nadap} = \texttt{GenerateAdaptedMeshes}(\{\mathcal{H}_{i}^{j-1}\}_{i=1,nadap}, \{\mathcal{M}_{i}^{j-1}\}_{i=1,nadap})$

```
End for
```

8.8. A numerical example: acoustic wave

A goal-oriented mesh adaptation algorithm is not designed to deliver an accurate or even simply convergent solution field since the adaptation effort is restricted to the best accuracy for the scalar output. If the scalar output for a transient simulation is a time integral, an interesting accuracy evaluation of a mesh-adaptive calculation concerns the *integrand* of this time integral. A successful mesh-adaptive calculation should show a third-order numerical convergence for a third-order accurate scheme. The experiment presented focus on the propagation of acoustic waves with an Euler model. It is a challenging simulation with a fixed but unstructured mesh for usual Euler schemes, which are generally too dissipative and dispersive. It is also a challenging problem for mesh adaptation since meshes cannot be regular and must follow traveling structures, which can go through the whole computational domain. We simulate the propagation of an acoustic wave between its emission and the time when the corresponding pressure fluctuation is recorded over a limited time interval with a microphone situated at some distance from the noise source. Not all sound waves will have influence on this record, and therefore, at each time level, only a part of the domain needs a fine mesh. And indeed, during these calculations, due to the effect of adjoint state, meshing concentrates only on a small part of the computational domain. We restrict to spatial accuracy and expect third-order convergence. In the two experiments which we describe, about 50 mesh-adaptation intervals (i.e. 50 meshes) are used in a time interval. The number of global fixed points iterations is 4. A typical wall time is 10 h on a laptop for the finest mesh.

The propagation considered is rather close to the one computed in section 7.7.1 of Chapter 7 of this volume, for which the Global Transient Fixed Point algorithm was combined with a second-order accurate scheme. For a sequence of adaptive meshes of 12K, 24K, 64K (mean size over the time interval), the numerical mesh convergence measured is 1.98. The challenge is now to get with the mesh adaptive algorithm a convergence numerical order close to 3. This computation is presented (in detail) in Carabias et al. (2018), in which unfortunately, because the computation was done on a not very fast laptop, the box is in this section two times smaller in order to have an easier case. We keep the same source term as [7.42] and the same output functional:

$$j(W) = \int_0^T \int_M \frac{1}{2} k(t) \, dM dt = \int_0^T \int_M \frac{1}{2} (p(\mathbf{x}, t) - p_0)^2 \, dM dt, \qquad [8.18]$$

where p_0 is the initial pressure. The mesh and pressure contours of two mesh-adaptative transient computations, a coarse one and a fine one are presented in Figure 8.4.

The numerical convergence order of the functional, is as expected, close to third order. But the method seems to provide also some superconvergence for some outputs. Indeed, let us analyze the convergence of the time-integrand $k(t)$ of the functional

$$k(t) = \int_M (p - p_0) dM$$

at some particular time $t = t^*$ for the different mesh-adaptive calculations. The variations of k are depicted in Figure 8.5. Numerical convergence order is measured on the maximum $k(t^*)$ of k, reached for time close to $t = 10.5$. It is obtained (by solving equation [6.3] (Chapter 6 of Volume 1) for three consecutive mesh sizes, the mean mesh sizes being successively 1,500, 5,400, 11,000 and 21,000. For the three finer calculations, with mean mesh sizes of 5,400, 11,000 and 21,000 vertices, convergence order appears as equal to 2.71, a level much higher than second-order case of section 7.7.1 (for which the fine adapted mesh was two times finer) and close to the theoretical order 3 (see Table 8.1). The meshes are anisotropic with aspect ratio larger than 3 to 4 in regions of interest. The smaller mesh size for the different meshes (at different times) are of order of 5×10^{-4}, which would be obtained with a uniform isotropic mesh of about 2 million vertices.

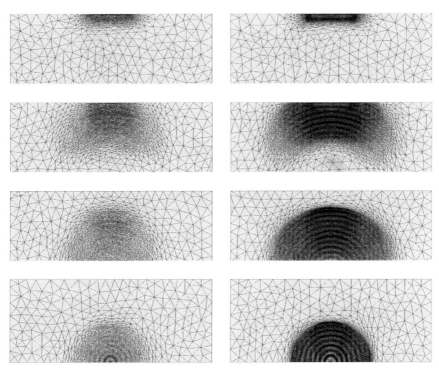

Figure 8.4. *Acoustic waves traveling in a box from bottom (5th mesh in time) to top (20th mesh in time) with the coarse option (left, 1,500 vertices in the mean) and the finest option (right, 21,566 vertices in the mean). For a color version of this figure, see www.iste.co.uk/dervieux/meshadaptation2*

Mean mesh size	Maximum close to T=10.5s	Observed convergence order
2,775	$6.893079.10^{-07}$	
5,403	$7.235717.10^{-07}$	
10,696	$7.431403.10^{-07}$	**1.70**
21,566	$7.510065.10^{-07}$	**2.71**

Table 8.1. *Noise propagation in a rectangular box: mesh convergence for the unsteady integrand*

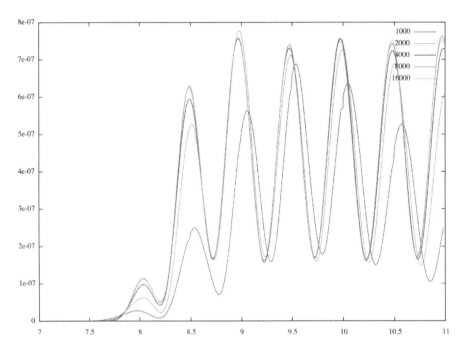

Figure 8.5. *Noise propagation in a box: mesh convergence of the spatial integral $k(t)$ (y-axis) as a function of time (x-axis). The lower curve "1000" corresponds to a mean mesh size of 1,500 vertices. The three other curves, almost undistinguishable, correspond to mean mesh sizes of 5,400, 11,000 and 21,000 vertices. For a color version of this figure, see www.iste.co.uk/dervieux/meshadaptation2*

8.9. Conclusion

This chapter describes an a priori analysis of the error of a third-order accurate CENO scheme for the unsteady Euler equations. The proposed analysis expresses all distributed error terms as functions of reconstruction errors. It allows to predict the effect of a given mesh (mesh size, anisotropy) on the approximation error. An important simplification consists in projecting the estimate on a metric-based analysis because of least square projection. This allows for an anisotropic mesh adaptation relying on an optimal metric, and solving goal-oriented problematics. A convergence order of 2.71, much better than with second-order accurate adaptation, is obtained with less degrees of freedom.

8.10. Notes

More computational results can be found in Carabias et al. (2018).

Let us also mention that the numerical scheme used here has been extended to shock capturing in Carabias (2013).

Short review of higher order adaptation. As already remarked in the chapter, interpolation errors are used for building adaptation criteria in Huang (2005). Several metrics are derived from the Hessians of each partial derivative. Then the metrics are intersected. A similar idea is discussed in Hecht (2008). Intersections of metrics do not produce really optimal meshes. Further they often result in loosing anisotropy. A true asymptotic extension is proposed in Cao (2005, 2008; see also Cao 2007a,b). We also refer to Mirebeau (2010) for similar ideas. A singular Sylvester decomposition is applied in Mbinky (2013). Further proposals for adapting a local molecule to a higher order polynomial are given in Coulaud and Loseille (2016). A proposal for h-p adaptation by measuring the different terms of the Taylor series is studied in Dolejsi (2014), Dolejsi et al. (2018), Rangarajan et al. (2017) (using a continuous mesh model) and Rangarajan et al. (2018). Researcher using CENO-like schemes have derived adapted approaches (see Ivan and Groth 2007; Pagnutti and Ollivier-Gooch 2009; Ivan 2011; Ivan et al. 2014). Approaches relying on optimization are proposed for higher order in Yano et al. (2011), Darmofal et al. (2013) and Yano and Darmofal (2014).

References

Abgrall, R. (1992). Design of an essentially non-oscillatory reconstruction procedure on finite-element type meshes. Technical report 1584, INRIA.

Abgrall, R. (2006). Residual distribution schemes: Current status and future trends. *Computers and Fluids*, 35, 641–669.

Absil, P.-A., Mahony, R., Sepulchre, R. (2008). *Optimization Algorithms on Matrix Manifolds*. Princeton University Press, Princeton, NJ.

Agouzal, A. and Vassilevski, Y.V. (2010). Minimization of gradient errors of piecewise linear interpolation on simplicial meshes. *Comp. Methods Appl. Mech. Eng.*, 199, 2195–2203.

Agouzal, A., Lipnikov, K., Vassilevski, Y. (2010). Hessian free metric based mesh adaptation via geometry of interpolation error. *Comput. Math. Math. Phys.*, 50(1), 124–138.

Alauzet, F. (2009). Size gradation control of anisotropic meshes. *Finite Elem. Anal. Des.*, 46, 181–202.

Alauzet, F. (2016). A parallel matrix-free conservative solution interpolation on unstructured tetrahedral meshes. *Comput. Methods Appl. Mech. Eng.*, 299, 116–142.

Alauzet, F. and Frazza, L. (2020). 3D RANS anisotropic mesh adaptation on the high-lift version of NASA's common research model (HL-CRM). AIAA Paper 2019-2947.

Alauzet, F. and Loseille, A. (2009a). High order sonic boom modeling by adaptive methods. RR-6845, INRIA.

Alauzet, F. and Loseille, A. (2009b). On the use of space filling curves for parallel anisotropic mesh adaptation. In *Proceedings of the 18th International Meshing Roundtable*, Clark, B.W. (ed.) Springer, Berlin.

Alauzet, F. and Loseille, A. (2010). High order sonic boom modeling by adaptive methods. *J. Comput. Phys.*, 229, 561–593.

Alauzet, F. and Mehrenberger, M. (2010). P1-conservative solution interpolation on unstructured triangular meshes. *Int. J. Numer. Methods Eng.*, 84(13), 1552–1588.

Alauzet, F. and Olivier, G. (2011). Extension of metric-based anisotropic mesh adaptation to time-dependent problems involving moving geometries. *49th AIAA Aerospace Sciences Meeting and Exhibit*, AIAA-2011-0896, Orlando, FL.

Alauzet, F., Frey, P.J., George, P.-L., Mohammadi, B. (2007). 3D transient fixed point mesh adaptation for time-dependent problems: Application to CFD simulations. *J. Comp. Phys.*, 222, 592–623.

Alauzet, F., Dervieux, A., Frazza, L., Loseille, A. (2019). Numerical uncertainties estimation and mitigation by mesh adaptation. In *Uncertainty Management for Robust Industrial Design in Aeronautics. Notes on Numerical Fluid Mechanics and Multidisciplinary Design, Vol. 66*, Hirsch, C., Wunsch, D., Szumbarski, J., Laniewski-Wollk, L., Pons-Prats, J. (eds). Springer, Dordrecht.

Anderson, W.K. and Venkatakrishnan, V. (1999). Aerodynamic design optimization on unstructured grids with a continuous adjoint formulation. *Comput. Fluids*, 28(4–5), 443–480.

Andrus, J.F. (1979). Numerical solution of systems of ordinary differential equations separated into subsystems. *SIAM J. Numer. Anal.*, 16(4), 605–611.

Andrus, J.F. (1993). Stability of a multi-rate method for numerical integration of ODEs. *Comput. Math. Appl.*, 25(2), 3–14.

Angrand, F.J. and Dervieux, A. (1984). Some explicit triangular finite element schemes for the Euler equations. *Int. J. Numer. Methods Fluids*, 4, 749–764.

Apel, T. (1999). *Anisotropic Finite Elements: Local Estimates and Applications*. Teubner, Stuttgart.

Arian, E. and Salas, M.D. (1999). Admitting the inadmissible: Adjoint formulation for incomplete cost functionals in aerodynamic optimization. *AIAA Journal*, 37(1), 37–44.

Arsigny, V., Fillard, P., Pennec, X., Ayache, N. (2006). Log-Euclidean metrics for fast and simple calculus on diffusion tensors. *Mag. Res. Med.*, 56(2), 411–421.

Babuška, I. and Strouboulis, T. (2001). *The Finite Element Method and its Reliability*. Oxford Scientific Publications, New York.

Baines, M. (1994). *Moving Finite Elements*. Oxford University Press, Inc., New York.

Bank, R.E. and Smith, R.K. (1993). A posteriori error estimate based on hierarchical bases. *SIAM J. Numer. Anal.*, 30, 921–935.

Barth, T.J. (1993). Recent developments of high-order k-exact reconstruction on unstructured meshes. *31st AIAA Aerospace Science Meeting*, AIAA-93-0668, Reno, NV.

Barth, T.J. and Frederickson, P.O. (1990). Higher order solution of the Euler equations on unstructured grids using quadratic reconstruction. AIAA-90-0013.

Barth, T.J. and Larson, M.G. (2002). A posteriori error estimation for higher order Godunov finite volume methods on unstructured meshes. In *Finite Volumes for Complex Applications III*, Herbin, R. and Kröner, D. (eds). Kogan Page Ltd, London.

Bassi, F. and Rebay, S. (1997). High-order accurate discontinuous finite element solution of the 2D Euler equations. *J. Comp. Phys.*, 138(2), 251–285.

Becker, R. and Rannacher, R. (1996). A feedback approach to error control in finite element methods: Basic analysis and examples. *East-West J. Numer. Math.*, 4, 237–264.

Becker, R., Braack, M., Rannacher, R. (1999). Numerical simulation of laminar flames at low Mach number with adaptative finite elements. *Combust. Theory Model.*, 3, 503–534.

Belme, A. (2011). Aérodynamique instationnaire et méthode adjointe. PhD Thesis, Université de Nice Sophia Antipolis, Sophia Antipolis [in French].

Belme, A., Dervieux, A., Alauzet, F. (2012). Time accurate anisotropic goal-oriented mesh adaptation for unsteady flows. *J. Comp. Phys.*, 231(19), 6323–6348.

Belme, A., Alauzet, F., Dervieux, A. (2019). An a priori anisotropic goal-oriented estimate for viscous compressible flow and application to mesh adaptation. *J. Comp. Phys.*, 376, 1051–1088.

Berger, M.J. and Colella, P. (1989). Local adaptive mesh refinement for shock hydrodynamics. *J. Comp. Phys.*, 82, 64–84.

Braack, R.B.M. and Rannacher, R. (1999). Numerical simulation of laminar flames at low Mach number with adaptative finite elements. *Combust. Theory Model.*, 3, 503–534.

Brèthes, G. and Dervieux, A. (2016). Anisotropic norm-oriented mesh adaptation for a Poisson problem. *J. Comp. Phys.*, 322, 804–826.

Brèthes, G. and Dervieux, A. (2017). A tensorial-based mesh adaptation for a Poisson problem. *Eur. J. Comput. Mech.*, 26(3), 245–281.

Brèthes, G., Allain, O., Dervieux, A. (2015). A mesh-adaptative metric-based full-multigrid for the Poisson problem. *Int. J. Numer. Methods Fluids*, 79(1) [Online]. Available at: http://www-sop.inria.fr/members/Gautier.Brethes/article-ADA-MG.pdf.

Bueno-Orovio, A., Castro, C., Palacios, F., Zuazua, E. (2012). Continuous adjoint approach for the Spalart–Allmaras model in aerodynamic optimization. *AIAA Journal*, 50(3), 631–646.

Cao, W. (2005). On the error of linear interpolation and the orientation, aspect ratio, and internal angles of a triangle. *SIAM J. Numer. Anal.*, 43(1), 19–40.

Cao, W. (2007a). Anisotropic measures of third order derivatives and the quadratic interpolation error on triangular elements. *SIAM J. Sci. Comput.*, 29(2), 756–781.

Cao, W. (2007b). An interpolation error estimate on anisotropic meshes in R^n and optimal metrics for mesh refinement. *SIAM J. Numer. Anal.*, 45(6), 2368–2391.

Cao, W. (2008). An interpolation error estimate in R^2 based on the anisotropic measures of higher derivatives. *Math. Comp.*, 77, 265–286.

Cao, W., Huang, W., Russell, R.D. (2003). Approaches for generating moving adaptive meshes: Location versus velocity. *Appl. Numer. Math.*, 47, 212–138.

Carabias, A. (2013). Analyse et adaptation de maillage pour des schémas non-oscillatoires d'ordre élevé. PhD Thesis, Université de Nice-Sophia-Antipolis, Nice [in French].

Carabias, A., Allain, O., Dervieux, A. (2011). Dissipation and dispersion control of a quadratic-reconstruction advection scheme. European Workshop on High Order Nonlinear Numerical Methods for Evolutionary PDEs: Theory and Applications, Trento.

Carabias, A., Belme, A., Loseille, A., Dervieux, A. (2018). Anisotropic goal-oriented error analysis for a third-order accurate CENO Euler discretization. *Int. J. Numer. Methods Fluids*, 86(6), 392–413.

Castro, C., Lozano, C., Palacios, F., Zuazua, E. (2007). Systematic continuous adjoint approach to viscous aerodynamic design on unstructured grids. *AIAA Journal*, 45(9), 2125–2139.

Charest, M.R.J., Groth, C.P.T., Gauthier, P.Q. (2015). A high-order central ENO finite-volume scheme for three-dimensional low-speed viscous flows on unstructured mesh. *Commun. Comput. Phys.*, 17(03), 615–656.

Ciarlet, P.G. and Raviart, P.A. (1972). General Lagrange and Hermite interpolation in R^n with applications to finite element methods. *Arch. Ration. Mech. Anal.*, 46, 177–199.

Clément, P. (1975). Approximation by finite element functions using local regularization. *Revue française d'automatique, informatique et recherche opérationnelle*, R-2, 77–84.

Cockburn, B. (2001). Devising discontinuous Galerkin methods for non-linear hyperbolic conservation laws. *J. Comput. Appl. Math.*, 128(1–2), 187–204.

Cockburn, B., Karniadakis, G., Shu, C.-W. (2000). *Discontinuous Galerkin Methods: Theory, Computation and Application*. Springer, Berlin.

Collins, M., Vecchio, F., Selby, R., Gupta, P. (1997). Failure of an offshore platform. *Concrete Int.*, 19(8), 28–35.

Constantinescu, E. and Sandu, A. (2007). Multirate timestepping methods for hyperbolic conservation laws. *J. Sci. Comp*, 33(3), 239–278.

Coulaud, O. and Loseille, A. (2016). Very high order anisotropic metric-based mesh adaptation in 3D. *Procedia Eng.*, 82, 353–365.

Courty, F., Leservoisier, D., George, P.-L., Dervieux, A. (2006). Continuous metrics and mesh adaptation. *Appl. Numer. Math.*, 56(2), 117–145.

Darmofal, D.L., Allmaras, S.R., Yano, M., Kudo, J. (2013). An adaptive, higher-order discontinuous Galerkin finite element method for aerodynamics. AIAA Conference Paper, AIAA 2013-2871, 1–23.

Derlaga, J.M. and Park, M.A. (2017). Application of exact error transport equations and adjoint error estimation to AIAA workshops. *55th AIAA Aerospace Sciences Meeting*, AIAA Paper 2017-0076, Grapevine, TX.

Dobrzynski, C. and Frey, P.J. (2008). Anisotropic Delaunay mesh adaptation for unsteady simulations. In *Proceedings of the 17th International Meshing Roundtable*, Garimella R.V. (ed.). Springer, Berlin.

Dolejsi, V. (2014). Anisotropic hp-adaptive method based on interpolation error estimates in the L^q-norm. *Appl. Numer. Math.*, 82, 80–114.

Dolejsi, V., May, G., Rangarajan, A. (2018). A continuous hp-mesh model for adaptive discontinuous Galerkin schemes. *Appl. Numer. Math.*, 124, 1–21.

Elias, R.N. and Coutinho, A.L.G.A. (2007). Stabilized edge-based finite element simulation of free-surface flows. *Int. J. Numer. Meth. Fluids*, 54(6–8), 965–993.

Engquist, S., Harten, B., Osher, A., Chakravarthy, S.R. (1986). Some results on uniformly high-order accurate essentially non-oscillatory schemes. *Appl. Numer. Math.*, 2(3–5), 347–377.

Engstler, C. and Lubich, C. (1997a). Multirate extrapolation methods for differential equations with different time scales. *Computing*, 58, 173–185.

Engstler, C. and Lubich, C. (1997b). Mur8: A multirate extension of the eight-order Dormer-Prince method. *Appl. Numer. Math.*, 25, 185–192.

Ern, A. and Vohralík, M. (2015). Polynomial-degree-robust a posteriori estimates in a unified setting for conforming, nonconforming, discontinuous Galerkin, and mixed discretizations. In *La Serena Numerica II, Octavo Encuentro de Anàlisis Numérico de Ecuaciones Diferenciales Parciales*. HAL Preprint 00921583.

Fedkiw, R., Aslam, T., Merriman, B., Osher, S. (1999). A non-oscillatory Eulerian approach to interfaces in multimaterial flows (the ghost fluid method). *J. Comput. Phys.*, 152(2), 457–492.

Feghaly, R., Raphael, W., Kaddah, F. (2008). Analyses of the reasons of Roissy terminal 2E collapse in France using deterministic and reliability assessments. Structures Congress, Vancouver.

Fidkowski, K.J. and Darmofal, D.L. (2011). Review of output-based error estimation and mesh adaptation in computational fluid dynamics. *AIAAJ*, 49(4), 673–694.

Formaggia, L. and Perotto, S. (2001). New anisotropic a priori error estimates. *Numer. Math.*, 89, 641–667.

Formaggia, L. and Perotto, S. (2003). Anisotropic a priori error estimates for elliptic problems. *Numer. Math.*, 94, 67–92.

Formaggia, L., Micheletti, S., Perotto, S. (2004), Anisotropic mesh adaptation in computational fluid dynamics: Application to the advection-diffusion-reaction and the Stokes problems. *Appl. Numer. Math.*, 51(4), 511–533.

Frazza, L., Loseille, A., Dervieux, A., Alauzet, F. (2019). Nonlinear corrector for Reynolds averaged Navier Stokes equations. *Int. J. Numer. Methods Fluids*, 91(11), 567–586.

Frey, P.J. (2001). YAMS, a fully automatic adaptive isotropic surface remeshing procedure. RT-0252, INRIA.

Gear, C.W. (1971). *Numerical Initial Value Problems in Ordinary Differential Equations*. Prentice Hall, Englewood Cliffs, NJ.

George, P.-L. (1999). Tet meshing: Construction, optimization and adaptation. In *Proceedings of the 8th International Meshing Roundtable*, Shimada, K. (ed.). South Lake Tao, CA.

Giles, M.B. (1987). Energy stability analysis of multi-step methods on unstructured meshes. CFDL Report 87-1, MIT Department of Aeronautics & Astronautics.

Giles, M.B. (1997a). On adjoint equations for error analysis and optimal grid adaptation in CFD. Technical Report NA-97/11, Oxford.

Giles, M.B. (1997b). Stability analysis of a Galerkin/Runge-Kutta Navier-Stokes discretisation on unstructured tetrahedral grids. *J. Comput. Phys.*, 132(2), 201–214.

Giles, M.B. and Pierce, N.A. (1999). Improved lift and drag estimates using adjoint Euler equations. AIAA Paper, 99-3293.

Giles, M.B. and Süli, A. (2002a). Adjoint methods for PDEs: A posteriori error analysis and postprocessing by duality. *Acta Numerica*, 11, 145–236.

Groth, C.P.T. and Ivan, L. (2011). High-order solution-adaptive central essentially non-oscillatory (CENO) method for viscous flows. *49th AIAA Aerospace Sciences Meeting including the New Horizons Forum and Aerospace Exposition*, AIAA 2011-367, January 4–7, Orlando, FL.

Guégan, D., Allain, O., Dervieux, A., Alauzet, F. (2010). An L^∞-L^p mesh adaptive method for computing unsteady bi-fluid flows. *Int. J. Numer. Methods Eng.*, 84(11), 1376–1406.

Günther, M. and Rentrop, P. (1993). Multirate row methods and latency of electric circuits. *Appl. Numer. Math.*, 13, 83–102.

Günther, M., Kvaerno, A., Rentrop, P., Guelhan, A., Klevanski, J., Willems, S. (1998). Multirate partitioned Runge-Kutta methods. *BIT*, 38(2), 101–104.

Günther, M., Kvaerno, A., Rentrop, P. (2001). Multirate partitioned Runge-Kutta methods. *BIT*, 41(3), 504–514.

Harten, A. and Chakravarthy, S. (1991). Multi-dimensional ENO schemes for general geometries. ICASE Report 91-76.

Hartman, R. (2008). Multitarget error estimation and adaptivity in aerodynamic flow simulations. *SIAM J. Sci. Comput.*, 31(1), 708–731.

Hay, A. and Visonneau, M. (2006). Error estimation using the error transport equation for finite-volume methods and arbitrary meshes. *Int. J. Comput. Fluid Dyn.*, 20(7), 463–479.

Hecht, F. (2008). Mesh generation and error indicator. Slides, CIRM Summer School: More Efficiency in Finite Element Methods, Valenciennes, September.

Hofer, E. (1976). A partially implicit method for large stiff systems of ODEs with only few equations introducing small time-constants. *SIAM J. Numer. Anal.*, 13(5), 645–666.

Huang, W. (2005). Metric tensors for anisotropic mesh generation. *J. Comp. Phys.*, 204(2), 633–665.

Huang, W. and Russel, R.D. (2011). *Adaptive Moving Mesh Methods*. Springer, Berlin.

Itam, E., Wornom, S., Koobus, B., Dervieux, A. (2018). Hybrid simulation of high-Reynolds number flows relying on a variational multiscale model. In *Progress in Hybrid RANS-LES Modelling, Notes on Numerical Fluid Mechanics and Multidisciplinary Design*, Hoarau, Y., Peng, S.-H., Schwamborn, D., Revell, D. (eds). Springer, Berlin.

Itam, E., Wornom, S., Koobus, B., Dervieux, A. (2019). A volume-agglomeration multirate time advancing for high Reynolds number flow simulation. *Int. J. Numer. Methods Fluids*, 89, 326–341.

Ivan, L. (2011). Development of high-order ceno finite-volume schemes with block-based adaptive mesh refinement. PhD Thesis, University of Toronto.

Ivan, L. and Groth, C.P.T (2007). High-order central ENO finite-volume scheme with adaptive mesh refinement, AIAA 2007-4323. *18th AIAA Computational Fluid Dynamics Conference*, June 25–28, Miami, FL.

Ivan, L. and Groth, C.P.T. (2014). High-order solution-adaptive central essentially non-oscillatory (CENO) method for viscous flows. *J. Comp. Phys.*, 257, 830–862.

Ivan, L., De Sterck, H., Groth, C.P.T. (2014). A fourth-order solution-adaptive CENO scheme for space-physics flows on three-dimensional multi-block cubed-sphere grids. *22nd Annual Conference of the CFD Society of Canada*, Toronto, Ontario.

Jakobsen, B. and Rosendahl, F. (1994). The Sleipner platform accident. *Structural Engineering International*, 4(3), 190–193.

Jones, W.T., Nielsen, E.J., Park, M.A. (2006). Validation of 3D adjoint based error estimation and mesh adaptation for sonic boom reduction. AIAA Paper, 2006–1150.

Karypis, G. and Kumar, V. (2006). A fast and high quality multilevel scheme for partitioning irregular graphs. *SIAM J. Sci. Comput.*, 20(1), 359–392.

Kirby, R. (2002). On the convergence of high resolution methods with multiple time scales for hyperbolic laws. *Math. Comput.*, 72(243), 129–1250.

Kleefsman, K.M.T., Fekken, G., Veldman, A.E.P., Iwanowski, B., Buchner, B. (2005). A volume-of-fluid based simulation method for wave impact problems. *J. Comput. Phys.*, 206(1), 363–393.

Koobus, B., Alauzet, F., Dervieux, A. (2011). Some compressible numerical models for unstructured meshes. In *CFD Hanbook*, Magoulès, F. (ed.). CRC Press, London.

Koshizuka, S., Tamako, S., Oka, Y. (1995). A particle method for incompressible viscous flows with fluid fragmentation. *Comput. Fluid Dyn. J.*, 4(1), 29–46.

Lafon, F.C. and Abgrall, R. (1993). ENO schemes on unstructured meshes. INRIA Report 2099.

Lallemand, M.H., Steve, H., Dervieux, A. (1992). Unstructured multigridding by volume agglomeration: Current status. *Comput. Fluids*, 21, 397–433.

Layton, W., Lee, H.K., Peterson, J. (2002). A defect-correction method for the incompressible Navier-Stokes equations. *Appl. Math. Comput.*, 129(1), 1–19.

Leicht, T. and Hartmann, R. (2010). Error estimation and anisotropic mesh refinement for 3D laminar aerodynamic flow simulations. *J. Comp. Phys.*, 229(19), 7344–7360.

Levy, W., Laflin, K.R., Tinoco, E.N., Vassberg, J.C., Mani, M., Rider, B., Morrison, J.H., Wahls, R.A., Morrison, J.H., Brodersen, O.P. et al. (2017). Summary of data from the fifth AIAA CFD drag prediction workshop. 51th AIAA Aerospace and Sciences Meeting, AIAA-2013-0046, Dallas, TX.

Löhner, R. (1989). Adaptive remeshing for transient problems. *Comput. Methods Appl. Mech. Eng.*, 75, 195–214.

Löhner, R., Morgan, K., Zienkiewicz, O.C. (1984). The use of domain splitting with an explicit hyperbolic solver. *Comput. Methods Appl. Mech. Eng.*, 45, 313–329.

Loseille, A. (2008). Adaptation de maillage anisotrope 3D multi-échelles et ciblée à une fonctionnelle pour la mécanique des fluides. Application à la prédiction haute-fidélité du bang sonique. PhD Thesis, Université Pierre et Marie Curie, Paris VI [in French].

Loseille, A. and Löhner, R. (2010). Adaptive anisotropic simulations in aerodynamics. 48th AIAA Aerospace Sciences Meeting and Exhibit, AIAA-2010-169, Orlando, FL.

Loseille, A. and Menier, V. (2013). Serial and parallel mesh modification through a unique cavity-based primitive. In *Proceedings of the 22th International Meshing Roundtable*, Sarrate, J. and Staten, M. (eds). Springer, Berlin.

Loseille, A., Dervieux, A., Frey, P.J., Alauzet, F. (2007). Achievement of global second-order mesh convergence for discontinuous flows with adapted unstructured meshes. AIAA Paper, 2007-4186.

Loseille, A., Dervieux, A., Alauzet, F. (2010a). A 3D goal-oriented anisotropic mesh adaptation applied to inviscid flows in aeronautics. *48th AIAA Aerospace Sciences Meeting and Exhibit*, AIAA-2010-1067, Orlando, FL.

Loseille, A., Dervieux, A., Alauzet, F. (2010b). Fully anisotropic goal-oriented mesh adaptation for 3D steady Euler equations. *J. Comput. Phys.*, 229, 2866–2897.

Loseille, A., Frazza, L., Alauzet, F. (2018). Comparing anisotropic adaptive strategies on the 2nd AIAA Sonic Boom Workshop geometry. *AIAA J.*, 56(3), 938–952.

Mbinky, E. (2013). Adaptation de maillages pour des interpolations d'ordre très élevé. Thesis, Pierre et Marie Curie, Paris VI.

Mbinky, E., Alauzet, F., Loseille, A. (2012). High-order interpolation for mesh adaptation. *Proceedings of ECCOMAS CFD*, Vienna.

Mer, K. (1998). Variational analysis of a mixed element volume scheme with fourth-order viscosity on general triangulations. *Comput. Methods Appl. Eng.*, 153, 45–62.

Michal, T., Babcock, D.S., Kamenetskiy, D., Krakos, J., Mani, M., Glasby, R.S., Erwin, T., Stefanski, D. (2017). Comparison of fixed and adaptive unstructured grid results for drag prediction workshop 6. AIAA Paper, 2017-0961. *55th AIAA Aerospace Sciences Meeting*, Grapevine, TX.

Michal, T., Kamenetskiy, D., Krakos, J. (2018a). Anisotropic adaptive mesh results for the third high lift prediction workshop (HiLiftPW-3). AIAA Paper, 2018-1257. *56th AIAA Aerospace Sciences Meeting*, Kissimmee, FL.

Michal, T., Kamenetskiyd, D.S., Marcum, D., Alauzet, F., Frazza, L., Loseille, A. (2018b). Comparing anisotropic error estimates for ONERA M6 wing RANS simulations. AIAA Paper, 2018-0920. *56th AIAA Aerospace Sciences Meeting*, Kissimmee, FL.

Mirebeau, J.-M. (2010). Optimal meshes for finite elements of arbitrary order. *Constr. Approx.*, 32, 339–383.

Mugg, P.R. (2012). Construction and analysis of multi-rate partitioned Runge-Kutta methods. Thesis, Naval Postgraduate School, Monterey, CA.

Oberkampf, W.L. and Trucano, T.G. (2002). Verification and validation in computational fluid dynamics. *Prog. Aerosp. Sci.*, 38(3), 209–272.

Ouvrard, H., Kozubskaya, T., Abalakin, I., Koobus, B., Dervieux, A. (2009). Advective vertex-centered reconstruction scheme on unstructured meshes. RR-7033, INRIA.

Pagnutti, D. and Ollivier-Gooch, C. (2009). A generalized framework for high order anisotropic mesh adaptation. *Comput. Struct.*, 87, 670–679.

Park, M.A. and Nemec, M. (2017). Near field summary and statistical analysis of the second AIAA Sonic Boom Prediction Workshop. 23th AIAA Computational Fluid Dynamics Conference, AIAA Paper 2017-3256, Denver, CO.

Picasso, M. (2003). An anisotropic error indicator based on Zienkiewicz-Zhu error estimator: Application to elliptic and parabolic problems. *SIAM J. Sci. Comp.*, 24(4), 1328–1355.

Pierce, N.A. and Giles, M.B. (2000). Adjoint recovery of superconvergent functionals from PDE approximations. *SIAM Review*, 42(2), 247–264.

Pierce, N.A. and Giles, M.B. (2004). Adjoint and defect error bounding and correction for functional estimates. *J. Comp. Phys.*, 200(2), 769–794.

Rangarajan, A.M., May, G., Dolejsi, V. (2017). Adjoint-based anisotropic mesh adaptation for discontinuous Galerkin methods using a continuous mesh model. *23rd AIAA Computational Fluid Dynamics Conference*. 10.2514/6.2017-3100.

Rangarajan, A.M., Balan, A., May, G. (2018). Mesh optimization for discontinuous Galerkin methods using a continuous mesh model. *AIAA J.*, 56(10), 4060–4073.

Rentrop, P. (1985). Partitioned Runge-Kutta methods with stiffness detection and step-size control. *Numer. Mathematik*, 47, 545–564.

Rice, J.R. (1960). Split Runge-Kutta method for simultaneous equations. *J. Res. Natl. Inst. Stand. Technol.*, 64B(3), 151–170.

Rogé, G. and Martin, L. (2008). Goal-oriented anisotropic grid adaptation / Adaptation de maillage anisotrope oriente objectif. *Comptes rendus mathematique*, 346(19–20), 1109–1112.

Rumsey, C.L. and Slotnick, J.P. (2015). Overview and summary of the second AIAA High Lift Prediction Workshop. *J. Aircraft*, 52(4), 1006–1025.

Rumsey, C.L., Slotnick, J.P., Long, M., Stuever, R.A., Wayman, T.R. (2011). Summary of the first AIAA CFD High-Lift Prediction Workshop. *J. Aircraft*, 48(6), 2068–2079.

Rumsey, C.L., Slotnick, J.P., Sclafani, A.J. (2018). Overview and summary of the third AIAA high lift prediction workshop. AIAA-Paper, 2018-1258. *56th AIAA Aerospace and Sciences Meeting*, AIAA-2018-1258, Kissimmee, FL.

Saad, Y. (2003). *Iterative Methods for Sparse Linear Systems*, 2nd edition. SIAM, Philadelpha, PA.

Sand, J. and Skelboe, S. (1992). Stability of backward Euler multirate methods and convergence of waveform relaxation. *BIT*, 32, 350–366.

Sandu, A. and Constantinescu, E. (2009). Multirate explicit Adams methods for time integration of conservation laws. *J. Sci. Comp.*, 38, 229–249.

Savcenco, V., Hundsdorfer, W., Verwer, J.G. (2007). A multirate time stepping strategy for stiff ordinary differential equations. *BIT*, 47, 137–155.

Seny, B., Lambrechts, J., Toulorge, T., Legat, V., Remacle, J.-F. (2014). An efficient parallel implementation of explicit multirate Runge-Kutta schemes for discontinuous Galerkin computations. *J. Comput. Phys.*, 256, 135–160.

Shu, C.W. and Cockburn, B. (2001). Runge-Kutta discontinuous Galerkin methods for convection-dominated problems. *J. Sci. Comput.*, 16(3), 173–261.

Shu, C.W. and Osher, S. (1988). Efficient implementation of essentially non-oscillatory shock-capturing schemes. *J. Comput. Phys.*, 77, 439–471.

Skelboe, S. (1989). Stability properties of backward differentiation multirate formulas. *Appl. Numer. Math.*, 5, 151–160.

Tinoco, E.N., Brodersen, O.P., Keye, S., Laflin, K.R., Feltrop, E., Vassberg, J.C., Mani, M., Rider, B., Wahls, R.A., Morrison, J.H. et al. (2017). Summary of data from the Sixth AIAA CFD Drag Prediction Workshop: CRM cases 2 to 5. *55th AIAA Aerospace and Sciences Meeting*, AIAA-2017-1208, Grapevine, TX.

Venditti, D.A. and Darmofal, D.L. (2000). Adjoint error estimation and grid adaptation for functional outputs: Application to quasi-one-dimensional flow. *J. Comput. Phys.*, 164(1), 204–227.

Venditti, D.A. and Darmofal, D.L. (2002). Grid adaptation for functional outputs: Application to two-dimensional inviscid flows. *J. Comput. Phys.*, 176(1), 40–69.

Venditti, D.A. and Darmofal, D.L. (2003). Anisotropic grid adaptation for functional outputs: Application to two-dimensional viscous flows. *J. Comput. Phys.*, 187(1), 22–46.

Verfürth, R. (1996). *A Review of A Posteriori Error Estimation and Adaptative Mesh-Refinement Techniques*. Wiley Teubner Mathematics, New York.

Verfürth, R. (2013). *A Posteriori Error Estimation Techniques for Finite Element Methods*. Oxford University Press, Oxford.

Weiner, R., Arnold, M., Rentrop, P., Strehmel, K. (1993). Partioning strategies in Runge-Kutta type methods. *IMA J. Numer. Analysis*, 13, 303–319.

Yan, G. and Olivier-Gooch, C.F. (2015). Accuracy of discretization error estimation by the error transport equation on unstructured meshes–Nonlinear systems of equations. 22nd AIAA Computational Fluid Dynamics Conference, AIAA Paper 2015-2747, Dallas, TX.

Yan, G. and Ollivier-Gooch, C.F. (2017). Towards higher order discretization error estimation by error transport using unstructured finite-volume methods for unsteady problems. *Comput. Fluids*, 154, 24.–255.

Yano, M. and Darmofal, D.L. (2014). Anisotropic simplex mesh adaptation by metric optimization for higher-order DG discretizations of 3D compressible flows. 10th WCCM, July 2012, Sao Paolo, Blucher Mechanical Engineering Proceedings, 1(1), 1–16.

Yano, M., Modisette, J.M., Darmofal, D.L. (2011). The importance of mesh adaptation for higher-order discretizations of aerodynamics flows. 20th AIAA Computational Fluid Dynamics Conference, AIAA-2011-3852, Honolulu, HI.

Zienkiewicz, O.C. and Zhu, J.Z. (1992). The superconvergent patch recovery and a posteriori error estimates. Part 1: The recovery technique. *Int. J. Numer. Meth. Eng*, 33(7), 1331–1364.

Index

A

a priori corrector, 7
acoustic waves, 134, 161–163, 165, 168, 183, 185, 186
algorithm 1.1 Finite volume corrector computation, 13, 14
algorithm 1.2 Finite element nonlinear corrector computation, 17
algorithm 2.2 Transient $\mathbf{L}^\infty(0, T; \mathbf{L}^p(\Omega))$ fixed-point mesh adaptation algorithm, 27, 36
algorithm 2.3 Convergent transient fixed-point mesh adaptation algorithm for flow with interface, 33, 34
algorithm 3.2 Multirate time-advancing, 49
algorithm 4.1 Adjoint-based mesh adaptation for a steady problem, 76, 90, 119
algorithm 5.1 Viscous goal-oriented mesh adaptation loop for steady flows, 90, 102, 104
algorithm 6.1 Norm-oriented mesh adaptation for a steady problem, 119, 120
algorithm 6.1 of Volume 1 Convergent steady fixed-point: Research of asymptotic convergence, 79, 102, 104
algorithm 7.1 Global transient fixed-point mesh adaptation algorithm, 159
algorithm 8.1 Third-order global fixed-point, 168, 183

anisotropic ratio/quotient of an element, 164
approximation ENO, 171
Aubin–Nitsche estimate, 86

B, C

blast wave, 134, 141, 143, 163–165
cell-agglomeration algorithm, 47
checkpointing, 140, 159, 182
code
    Aironum, 55
    NiceFlow, 33, 36
    Wolf, 77, 79, 102, 161
convergence order, 23
corrector system, 4, 6, 9, 11, 14, 121, 122, 127, 130

D, E

decimation, 78
defect-correction (DC) corrector, 9
dispersion error, dissipation error, 174, 183
early capturing (EC) property of an adaptation, 22, 104
equidistribution adaption method, 155

F, G

feature-based approach, 21, 65, 77, 104, 105, 111, 114, 119, 124, 125, 127, 128, 131, 133 goal-oriented adaptation method, 3, 18, 19, 65, 66, 68, 69, 76, 77, 79, 81, 83, 111–113, 116, 119, 126, 131–133, 146, 148, 155, 158, 160, 161, 168, 171

L, M

level set method, 21, 22, 30
median finite volume cell, 47, 171
multiscale mesh adaptation, 21

MUSCL upwind scheme, 45, 62

N, R, S

numerical convergence order, 184
recovery methods, 7, 78, 103, 161
Spalart–Allmaras turbulence model, 9, 10, 129
superconvergent scheme, 55

Summary of Volume 1

Acknowledgments

Introduction

Chapter 1. CFD Numerical Models

1.1. Compressible flow
   1.1.1. Introduction
   1.1.2. Spatial representation
   1.1.3. Spatial second-order accuracy: MUSCL
   1.1.4. Low dissipation advection schemes
   1.1.5. Time advancing
   1.1.6. Positivity of mixed element-volume formulations
1.2. Viscous compressible flows
   1.2.1. Model for laminar flows
   1.2.2. Boundary conditions spatial discretization
   1.2.3. No-slip boundary condition
   1.2.4. Slip boundary condition
   1.2.5. Influence stencil
   1.2.6. Spalart–Allmaras one equation turbulence model
   1.2.7. SA one-equation model without trip and without f_{t2} term
   1.2.8. "Standard" SA one-equation model (without trip)
   1.2.9. "Full" SA one-equation model (with trip)
   1.2.10. Mixed element-volume discretization of SA
   1.2.11. Implicit time integration
1.3. A multi-fluid incompressible model
   1.3.1. Introduction
   1.3.2. Bi-fluid incompressible Navier–Stokes equations
   1.3.3. Finite element approximation

1.3.4. Error estimate for the level set advection
1.3.5. Provisional conclusion on scheme accuracy
1.4. Appendix: circumcenter cells
1.4.1. Two-dimensional circumcenter cells
1.4.2. Three-dimensional circumcenter cells
1.5. Notes

Chapter 2. Mesh Convergence and Barriers

2.1. Introduction
2.2. The early capturing property
2.2.1. Smoothness, non-smoothness, heterogeneity
2.2.2. Behavior of the uniform-mesh strategy
2.2.3. An example of 1D adaptation
2.3. Unstructured meshes in finite element method
2.3.1. Basics of finite element meshes
2.3.2. Anisotropy
2.4. Accuracy of an interpolation
2.5. Isotropic adaptative interpolation
2.5.1. The 2D case
2.5.2. A first 3D case
2.5.3. A limiting barrier for the isotropic 3D case
2.6. Anisotropic adaptative interpolation
2.6.1. Anisotropic adaptation of a Heaviside function
2.6.2. Heaviside function with curved discontinuity
2.7. Numerical illustration: anisotropic versus isotropic interpolation
2.8. CFD applications of anisotropic capture
2.8.1. Pressure with discontinuous gradient
2.8.2. Scramjet flow
2.9. Unsteady case
2.9.1. Barriers for second-order time-leveled case
2.9.2. Barriers for third-order time-leveled case
2.10. Conclusion
2.11. Notes

Chapter 3. Mesh Representation

3.1. Introduction
3.2. An introductory example
3.3. Euclidean metric space
3.3.1. Geometric interpretation
3.3.2. Natural metric mapping
3.4. Riemannian metric space
3.5. Generation of adapted anisotropic meshes

3.5.1. Unit element
3.5.2. Geometric invariants
3.5.3. Global duality
3.5.4. Quantifying mesh anisotropy
3.6. Operations on metrics
3.6.1. Metric intersection
3.6.2. Metric interpolation
3.7. Computation of geometric quantities
3.7.1. Computation of lengths
3.7.2. Computation of volumes
3.8. Notes
3.8.1. A short history

Chapter 4. Geometric Error Estimate
4.1. The 1D case
4.1.1. 1D metric
4.1.2. P^1 Interpolation error bound
4.1.3. 1D optimal metric
4.1.4. Convergence order of the continuous metric model
4.2. Discrete-continuous duality for linear interpolation error
4.2.1. Interpolation error in $\mathbf{L}^1$ norm for quadratic functions
4.2.2. Linear interpolation on a continuous element
4.2.3. Continuous linear interpolate
4.3. Numerical validation of the continuous interpolation error
4.3.1. Continuous interpolation error calculation
4.3.2. Comparison with discrete interpolation error computation
4.3.3. Three-dimensional validation
4.3.4. Some conclusions
4.4. Optimal control of the interpolation error in $\mathbf{L}^p$ norm
4.4.1. Formal resolution
4.4.2. Uniqueness
4.4.3. Optimal orientations and main result
4.5. Multidimensional discontinuity capturing
4.6. Linear interpolate operator
4.7. A local $\mathbf{L}^\infty$ upper bound of the interpolation error
4.8. Metric construction for mesh adaptation
4.8.1. Handling degenerated cases
4.8.2. Isotropic mesh adaptation
4.9. Mesh adaptation for analytical functions
4.9.1. Algorithms
4.9.2. Examples of L^∞ adaptation
4.10. Conclusion
4.11. Notes

Chapter 5. Multiscale Adaptation for Steady Simulations
5.1. Introduction
5.2. Definitions and notations (2D)
5.3. Solving the problematic of the unknown solution (2D/3D)
5.4. Numerical computation/recovery of the Hessian matrix
    5.4.1. Numerical computation of nodal gradients (2D)
    5.4.2. A double $\mathbf{L}^2$-projection method
    5.4.3. A method based on the Green formula
    5.4.4. A least-square approach
    5.4.5. From our experience
    5.4.6. Discrete-continuous interpolation
5.5. Solution interpolation
    5.5.1. Localization algorithm
    5.5.2. Classical polynomial interpolation
5.6. Mesh adaptation algorithm
5.7. Example of a CFD numerical simulation
5.8. Conclusion
5.9. Notes
    5.9.1. A short review of mesh/PDE coupling

Chapter 6. Multiscale Convergence and Certification in CFD
6.1. Introduction
6.2. A mesh convergence algorithm
    6.2.1. Mesh adaptation with a fixed complexity
    6.2.2. Transfers and numerical convergence
6.3. An academic test case
    6.3.1. Uniform refinement study
    6.3.2. Isotropic adaptation study
    6.3.3. Anisotropic adaptation study
    6.3.4. Error level
6.4. 3D multiscale anisotropic mesh adaptation
6.5. Conclusion
6.6. Notes

References

Other titles from

in

Numerical Methods in Engineering

2022

BOUCLIER Robin, HIRSCHLER Thibault
IGA: Non-conforming Coupling and Shape Optimization of Complex Multipatch Structures (Isogeometric Analysis Tools for Optimization Applications in Structural Mechanics Set – Volume 1)

GRANGE Stéphane, SALCIARINI Diana
Deterministic Numerical Modeling of Soil–Structure Interaction

2021

GENTIL Christian, GOUATY Gilles, SOKOLOV Dmitry
*Geometric Modeling of Fractal Forms for CAD
(Geometric Modeling and Applications Set – Volume 5)*

2020

GEORGE Paul Louis, ALAUZET Frédéric, LOSEILLE Adrien, MARÉCHAL Loïc
*Meshing, Geometric Modeling and Numerical Simulation 3: Storage, Visualization and In Memory Strategies
(Geometric Modeling and Applications Set – Volume 4)*

SIGRIST Jean-François
Numerical Simulation, An Art of Prediction 2: Examples

2019

DA Daicong
Topology Optimization Design of Heterogeneous Materials and Structures

GEORGE Paul Louis, BOROUCHAKI Houman, ALAUZET Frédéric, LAUG Patrick, LOSEILLE Adrien, MARÉCHAL Loïc
Meshing, Geometric Modeling and Numerical Simulation 2: Metrics, Meshes and Mesh Adaptation
(Geometric Modeling and Applications Set – Volume 2)

MARI Jean-Luc, HÉTROY-WHEELER Franck, SUBSOL Gérard
Geometric and Topological Mesh Feature Extraction for 3D Shape Analysis
(Geometric Modeling and Applications Set – Volume 3)

SIGRIST Jean-François
Numerical Simulation, An Art of Prediction 1: Theory

2017

BOROUCHAKI Houman, GEORGE Paul Louis
Meshing, Geometric Modeling and Numerical Simulation 1: Form Functions, Triangulations and Geometric Modeling
(Geometric Modeling and Applications Set – Volume 1)

2016

KERN Michel
Numerical Methods for Inverse Problems

ZHANG Weihong, WAN Min
Milling Simulation: Metal Milling Mechanics, Dynamics and Clamping Principles

2015

ANDRÉ Damien, CHARLES Jean-Luc, IORDANOFF Ivan
3D Discrete Element Workbench for Highly Dynamic Thermo-mechanical Analysis
(Discrete Element Model and Simulation of Continuous Materials Behavior Set – Volume 3)

JEBAHI Mohamed, ANDRÉ Damien, TERREROS Inigo, IORDANOFF Ivan
Discrete Element Method to Model 3D Continuous Materials
(Discrete Element Model and Simulation of Continuous Materials Behavior Set – Volume 1)

JEBAHI Mohamed, DAU Frédéric, CHARLES Jean-Luc, IORDANOFF Ivan
Discrete-continuum Coupling Method to Simulate Highly Dynamic Multi-scale Problems: Simulation of Laser-induced Damage in Silica Glass
(Discrete Element Model and Simulation of Continuous Materials Behavior Set – Volume 2)

SOUZA DE CURSI Eduardo
Variational Methods for Engineers with Matlab®

2014

BECKERS Benoit, BECKERS Pierre
Reconciliation of Geometry and Perception in Radiation Physics

BERGHEAU Jean-Michel
Thermomechanical Industrial Processes: Modeling and Numerical Simulation

BONNEAU Dominique, FATU Aurelian, SOUCHET Dominique
Hydrodynamic Bearings – Volume 1
Mixed Lubrication in Hydrodynamic Bearings – Volume 2
Thermo-hydrodynamic Lubrication in Hydrodynamic Bearings – Volume 3
Internal Combustion Engine Bearings Lubrication in Hydrodynamic Bearings – Volume 4

DESCAMPS Benoît
Computational Design of Lightweight Structures: Form Finding and Optimization

2013

YASTREBOV Vladislav A.
Numerical Methods in Contact Mechanics

2012

DHATT Gouri, LEFRANÇOIS Emmanuel, TOUZOT Gilbert
Finite Element Method

SAGUET Pierre
Numerical Analysis in Electromagnetics

SAANOUNI Khemais
Damage Mechanics in Metal Forming: Advanced Modeling and Numerical Simulation

2011

CHINESTA Francisco, CESCOTTO Serge, CUETO Elias, LORONG Philippe
Natural Element Method for the Simulation of Structures and Processes

DAVIM Paulo J.
Finite Element Method in Manufacturing Processes

POMMIER Sylvie, GRAVOUIL Anthony, MOËS Nicolas, COMBESCURE Alain
Extended Finite Element Method for Crack Propagation

2010

SOUZA DE CURSI Eduardo, SAMPAIO Rubens
Modeling and Convexity

2008

BERGHEAU Jean-Michel, FORTUNIER Roland
Finite Element Simulation of Heat Transfer

EYMARD Robert
Finite Volumes for Complex Applications V: Problems and Perspectives

FREY Pascal, GEORGE Paul Louis
Mesh Generation: Application to finite elements – 2nd edition

GAY Daniel, GAMBELIN Jacques
Modeling and Dimensioning of Structures

MEUNIER Gérard
The Finite Element Method for Electromagnetic Modeling

2005

BENKHALDOUN Fayssal, OUAZAR Driss, RAGHAY Said
Finite Volumes for Complex Applications IV: Problems and Perspectives

Printed and bound by CPI Group (UK) Ltd, Croydon, CR0 4YY
05/10/2022
03152175-0001